Manfred Knebusch
Manfred Kolster

Wittrings

Aspects of Mathematics
Aspekte der Mathematik

Editor: Klas Diederich

vol. E 1: G. Hector/U. Hirsch, Introduction to the Geometry of Foliations, Part A
vol. E 2: M. Knebusch/M. Kolster, Wittrings

The text published in this series are intended for graduate students and all mathematicians who wish to broaden their research horizons or who simply want to get a better idea of what is going in a given field. They are introductions to areas close to modern research at a high level and prepare the reader for a better understanding of research papers. Many of the books can also be used to supplement graduate course programs.

The series will comprise two sub-series, one with English texts only and the other in German.

Manfred Knebusch
Manfred Kolster

Wittrings

V

Friedr. Vieweg & Sohn Braunschweig/Wiesbaden

CIP-Kurztitelaufnahme der Deutschen Bibliothek

Knebusch, Manfred:
Wittrings/Manfred Knebusch; Manfred Kolster. –
Braunschweig; Wiesbaden: Vieweg, 1982.
(Aspects of mathematics; Vol. 2)

NE: Kolster, Manfred; GT

Dr. *Manfred Knebusch* is Full Professor at the University of Regensburg, Germany.

Dr. *Manfred Kolster* is Assistant at the University of Münster, Germany.

Produced by Lengericher Handelsdruckerei, Lengerich

ISBN-13: 978-3-528-08512-4 e-ISBN-13: 978-3-322-84382-1
DOI: 10.1007/978-3-322-84382-1

Preface

This book is intended to give an introduction to some parts of the algebraic theory of quadratic forms. It grew out of a course, given by the first author at the University of Saarbrücken during the academic year 1971/72. This course considered "weak methods" in the theory of quadratic forms, i.e. methods, which apply to the classes of forms in the Wittring rather than to the forms themselves. In the meantime some topics of this course (cf. VIII) have been treated in an excellent way in other books - we only mention Lam [41].

The part of the course covered by the present book is the ringtheoretic approach to the structure of Wittrings in the framework of "abstract" Wittrings, based on the papers [36], [37]. Working with these rings yields a greater flexibility than just working with the Wittrings of fields. Besides this the study of abstract Wittrings has the advantage, that the results are not limited to the case of Wittrings over fields, but carry over to local rings, and thus give an indespensable contribution to the local part of a theory of forms over algebraic varieties (cf. [39] for a first impression of this theory). After having read this book, the reader will be well-prepared to understand the local theory (cf. e.g. [39]).

The first chapter of the book presents a brief account of Witt's theory of symmetric bilinear forms over fields and serves as a motivation for the introduction of abstract Wittrings. The ring-theoretic properties of such rings - mainly found by Pfister [52] in the case of Wittrings over fields - are considered in the second chapter. Finally, the third chapter deals with reduced abstract Wittrings as subrings of rings of continuous integral - valued functions on a Boolean space. The reduced Wittrings of real fields have been studied intensively throughout the last decade, mainly because of their connections with the orderings of fields, and are now fairly well understood (cf. Becker-Bröcker [10]). Some of the main results of this theory have been generalized by Marshall ([45], [46], [47], [48]) in an abstract setting, that fits well into the approach chosen in this book.

We tried to incorporate into this book only topics, which are absolutely necessary to understand the modern theory of quadratic forms over fields and more generally over algebraic varieties, to give the reader a "cheap" entrance to this now rapidly developing part of mathematics. We feel, that this book would serve it's purpose, if it could stimulate and enable the reader to go further and deeper into the theory.

Regensburg, Münster, Juli 1981

Manfred Knebusch, Manfred Kolster

Contents

Vorwort*)

von Manfred Knebusch

Im Sommersemester 1971 und Wintersemester 1971/72 hielt ich an der Universität des Saarlandes eine zweistündige Vorlesung über die algebraische Theorie der quadratischen Formen über Körpern. Das Ziel der Vorlesung war, eine Einführung in die sogenannten "schwachen Methoden" (vgl. Scharlau, J.Number Theory 4, p.80 ff) zu geben, bei denen man weniger mit den Formen selbst als mit ihren Witt-Klassen, d.h. ihren Bildern im Wittring W(F) des Grundkörpers F, arbeitet. Genauer hatte die Vorlesung folgenden Inhalt:

1) Die von A.Rosenberg, R.Ware und mir entwickelte "ringtheoretische" Methode (s. [36], [37]).
2) Die von W.Scharlau entwickelte "Verlagerungsmethode" (s. [58]).
3) Anwendung auf die Theorie der reellen Abschlüsse von Körpern (s. [35], [9]).

Herr Kolster, jetzt Universität Münster, hat die Vorlesung sorgfältig ausgearbeitet und überdies den Text wegen des starken Interesses auf der anderen Seite des großen Teiches ins Englische übertragen. Das Manuskript schlummerte dann fast fünf Jahre in einer Schublade. Inzwischen gab es eine rasche Weiterentwicklung der alge-

*) This is the preface of the first edition as Regensburger Trichter 14 (1978).

braischen Theorie der quadratischen Formen, und uns war unklar, wie weit es von Nutzen war das Manuskript zu veröffentlichen. Mancherlei aus der Vorlesung findet man inzwischen gut dargestellt in anderen Büchern, s. insbesondere T.Y.Lam's Buch [41]. Anderes, wie etwa die von A.Dress studierte Amitsur-Kohomologie der Wittringe [30], hat bisher wenig Anwendung gefunden. Deshalb habe ich den Text von Kolster überarbeitet und um etwa die Hälfte des Inhaltes gekürzt, so daß im wesentlichen nur die ringtheoretische Methode zur Sprache kommt. Der damit verbliebene hier vorgelegte Restbestand scheint uns auch heute noch eine nützliche, leichte und schnelle Einführung in gewisse Aspekte der algebraischen Theorie der quadratischen Formen zu sein. 1)

Zweifellos liefern die "starken Methoden" wie sie insbesondere von A.Pfister entwickelt wurden (Teilformensatz, stark multiplikative Formen, u.s.w., s. Lorenz [44] und das oben zitierte Buch [41] von Lam) Einblicke in die Theorie der quadratischen Formen über Körpern, die uns hier versagt bleiben. Ich rate deshalb dem Neuling dringend, sich auch mit den starken Methoden vertraut zu machen. Neben den Büchern von Lorenz und Lam empfehle ich für einen Überblick darüber meine Arbeiten "Generic splitting of quadratic forms I und II Proc. London Math. Soc. 33 (1976) und 34 (1977). Als allererste Einführung in die algebraische Theorie der quadratischen Formen überhaupt empfehle ich die Vorlesung von Lam [42] auf der Konferenz über

1) Es wurden zwei Paragraphen (III, § 4, § 5) hinzugefügt, deren Inhalt 1971 noch nicht bekannt war.

quadratische Formen in Kingston 1976. Wohlverstanden gibt es auch eine - tiefgründige - arithmetische und analytische Theorie der quadratischen Formen (Fermat, Euler, Lagrange, ..., Hasse, Siegel, Hecke, ...), mit der wir hier aber nichts zu schaffen haben.

Bei der hier dargestellten ringtheoretischen Methode werden "abstrakte Wittringe" studiert. Das ist eine Klasse von kommutativen Ringen mit Einselement, welche die Wittringe W(F) umfaßt, aber eine größere algebraische Beweglichkeit gestattet als die Kategorie der Ringe W(F) selbst. Zum Beispiel ist das Bild von W(F) unter einem beliebigen Ringhomomorphismus in einen Wittring W(F') stets ein abstrakter Wittring, ohne notwendig zu dem Wittring eines Körpers isomorph zu sein. Ebenso ist das Tensorprodukt $W(F_1) \otimes_{\mathbb{Z}} W(F_2)$ zu zwei Körpern F_1, F_2 und allgemeiner jedes Tensorprodukt $W(F_1) \otimes_{W(F)} W(F_2)$ ein abstrakter Wittring. In der Arbeit [36] werden noch allgemeinere abstrakte Wittringe als in dem jetzigen Text studiert. Doch haben diese Ringe bisher keine Anwendungen in der Theorie der quadratischen Formen gefunden.

Warum ist es auch heute eine sinnvolle Möglichkeit die algebraische Theorie der quadratischen Formen mit einem Studium abstrakter Wittringe zu beginnen? Neben der soeben angeführten größeren "algebraischen Beweglichkeit" gibt es dafür einen weiteren - wichtigeren - Grund. Man kann symmetrische Bilinearformen über algebraischen Mannigfaltigkeiten anstelle von Körpern studieren (s. dazu meine Vorlesung auf der Konferenz über quadratische Formen in Kingston 1976, [39]). Der lokale Teil dieser Theorie hat als Gegenstand den Wittring W(A) eines kommutativen lokalen Ringes A, dessen Elemente Klassen von symmetrischen Bilinearformen über A sind. Diese Ringe W(A) erweisen sich als abstrakte Wittringe in unserem Sinne, und es läßt sich fast alles, was auf den nächsten Seiten über Körpern bewiesen wird, auf lokale Ringe übertragen. Bisher ist es aber nicht gelungen, die starken Methoden von Pfister bei lokalen Ringen anzuwenden. Für einen ersten Eindruck von dieser "lokalen algebraischen Theorie" empfehle ich Kapitel II meiner Kingston Vorlesung 1976 ([39]).

Darüber hinaus gibt es andere in Verbindung mit quadratischen und bilinearen Formen auftretende Ringe, die abstrakte Wittringe sind, etwa die von Scharlau [56] und Belskij [11] studierten "kohomologischen Wittringe".

Regensburg, September 1978

M.Kn.

Chapter I: Basic facts about symmetric bilinear forms, and definition of the Witt ring.

In this introductory chapter we give a brief account of some aspects in the classical theory of symmetric bilinear spaces over fields. For a more detailed discussion the reader should consult [12],[49],[59],[41].We will often restrict ourselves to fields with characteristic different from two. Remarks concerning peculiarities of the characteristic two case will be marked by an asterisk.

§ 1 Bilinear spaces

Let F be a field. A (symmetric) bilinear space over F is a pair (E,B), where E is a finite dimensional vectorspace over F and $B : E \times E \to F$ is a symmetric bilinear form on E, i.e. B is biadditive $B(cx,y) = c\cdot B(x,y)$ and $B(x,y) = B(y,x)$ for all x,y in E and c in F. If the bilinear form B is clear from the context we simply write E for the bilinear space (E,B). Every subspace W of E endowed with the restriction of B to $W \times W$ is a bilinear space. Two vectors x and y in a bilinear space (E,B) are called orthogonal if $B(x,y) = 0$. If W is a subspace of E, we define $W^\perp := \{x \in E;\ B(x,W) = 0\}$. Here $B(x,W) =$ means that x and y are orthogonal for all y in W. The set $W^\perp$ is actually a subspace of E, called the orthogonal complement of W in E. $E^\perp$ is the radical of E. We call E non singular if $E^\perp = 0$.

Let $\{e_1,\ldots,e_n\}$ be a basis of E over F. The bilinear form B is uniquely determined by the symmetric $n \times n$ - matrix $\mathfrak{B} :=$ $(B(e_i,e_j))$. Let us take another basis $\{f_1,\ldots,f_n\}$ of E and set $\mathfrak{B}' :=$

$(B(f_i,f_j))$. Then we have $\mathfrak{B}' = {}^t\mathfrak{A}\,\mathfrak{B}\,\mathfrak{A}$, where $\mathfrak{A} := (a_{ij})$ comes from the change of basis $f_j = \sum_{i=1}^{n} e_i\, a_{ij}$ for $j = 1,\ldots,n$. We see that $\det \mathfrak{B}' = (\det \mathfrak{A})^2 \det \mathfrak{B}$. Thus the class of $\det \mathfrak{B}$ in $F^*/_{F^{*2}} \cup \{0\}$ – we denote by F^* the group of elements $\neq 0$ of F – is independent of the basis chosen for E. This invariant of the bilinear space (E,B) is called the <u>determinant</u> of (E,B) and written $\det E$ or $\det B$. A linear map $s : (E,B) \to (E',B')$ between bilinear spaces over F is called an <u>isometry</u> or <u>isomorphism</u> if s is an isomorphism between the underlying vector spaces and $B'(sx,sy) = B(x,y)$ for all x,y in E. Clearly (E,B) and (E',B') are isomorphic if and only if $\dim E = \dim E'$ and $\mathfrak{B}' = {}^t\mathfrak{A}\,\mathfrak{B}\,\mathfrak{A}$ where $\mathfrak{B}$ resp. $\mathfrak{B}'$ belong to B resp. B' with respect to some basis of E resp. E' and $\mathfrak{A}$ is some $n \times n$ - matrix over F with $n := \dim E = \dim E'$.

In the following we sometimes use matrix notation to describe a bilinear space. For instance a one-dimensional space (Fz,B) with $B(z,z) = a$ will be written (a) for short. We know from above that $(a) \cong (b)$ if and only if $a = c^2 b$ for some c in F^*.

Let E^* denote the dual space $\mathrm{Hom}_F(E,F)$ of E. The bilinear form B induces a homomorphism $\varphi_B : E \to E^*$ given by $\varphi_B(x) = B(\cdot,x)$. Obviously we get

<u>Prop. 1.1.</u> (E,B) is non singular $\Leftrightarrow$ φ_B is injective $\Leftrightarrow$ φ_B is bijective $\Leftrightarrow$ $\det B \neq 0$.

Let (E_1,B_1) and (E_2,B_2) be bilinear spaces over F. We equip the direct sum $E_1 \oplus E_2$ with a bilinear form B defined by $B(e_1 \oplus e_2,\ f_1 \oplus f_2) := B_1(e_1,f_1) + B_2(e_2,f_2)$ for all e_1,f_1 in E_1 and e_2,f_2 in E_2. The space $(E_1 \oplus E_2,B)$ is called the <u>orthogonal sum</u> of

(E_1,B_1) and (E_2,B_2) and is written $(E_1,B_1) \perp (E_2,B_2)$ or $E_1 \perp E_2$. If we identify the dual space $(E_1 \oplus E_2)^*$ with the sum $E_1^* \oplus E_2^*$ in the usual way, we get $\varphi_B = \varphi_{B_1} \oplus \varphi_{B_2}$. Using Prop. 1.1 we see that $E_1 \perp E_2$ is non singular if and only if E_1 and E_2 are non singular. If the matrix $\mathfrak{B}_1$ represents (E_1,B_1) with respect to some basis, and if the matrix $\mathfrak{B}_2$ represents (E_2,B_2) with respect to some basis, the orthogonal sum $E_1 \perp E_2$ is represented by the matrix $\begin{pmatrix} \mathfrak{B}_1 & 0 \\ 0 & \mathfrak{B}_2 \end{pmatrix}$ with respect to the composition of the two bases. This implies at once that $\det(E_1 \perp E_2) = \det E_1 \cdot \det E_2$. If we have two subspaces F_1 and F_2 of a bilinear space (E,B) such that $E = F_1 \oplus F_2$ and $B(F_1,F_2) = 0$, we also write $E = F_1 \perp F_2$, since evidently (E,B) is canonically isomorphic to $(F_1,B|F_1 \times F_1) \perp (F_2,B|F_2 \times F_2)$.

Every decomposition $E = W \oplus E^{\perp}$ is orthogonal. $(W,B|W \times W)$ is uniquely determined up to isomorphism, since B obviously induces a bilinear form $\overline{B}$ on the vector space $\overline{E} := E/E^{\perp}$ and the canonical map $W \to \overline{E}$ is an isometry with respect to $B|W \times W$ and $\overline{B}$. In the following we shall therefore consider only non singular spaces, if we do not explicitly allow singular ones.

<u>Prop. 1.2.</u> Let (E,B) be a (not necessarily non singular) bilinear space over F, and let W be a non singular subspace of E. Then $E = W \perp W^{\perp}$.

<u>Proof.</u> Denote $B|W \times W$ by B_o. Since W is non singular, we can use Prop. 1.1. This gives us for every x in E a uniquely determined element $\pi(x)$ in W, such that $\varphi_B(x)|W = \varphi_{B_o}(\pi(x))$. Thus the map $\pi : E \to W$ is a well-defined projection from E onto W. We have $E = W \oplus \ker \pi$, but $\ker \pi$ is obviously equal to $W^{\perp}$.

Let x be an element of a space (E,B). We define the <u>norm</u> of x to be $n(x) := B(x,x)$. x is called <u>anisotropic</u>, if $n(x) \neq 0$. Otherwise it is called <u>isotropic</u>. The space E is <u>anisotropic</u>, if each element $\neq 0$ in E is anisotropic, otherwise E is <u>isotropic</u>. A subspace W of E is called <u>totally isotropic</u> if $W \subset W^{\perp}$. This means that the form B restricted to $W \times W$ is identically zero.

<u>Theorem 1.3.</u> Let char $F \neq 2$. Every bilinear space over F has an orthogonal decomposition into one-dimensional spaces.

<u>Proof.</u> i) As a first step let us prove that a bilinear space (E,B) contains anisotropic vectors. Take $x \neq 0$ in E. If $n(x) \neq 0$, we are done. Otherwise there is at least one y in E with $B(x,y) \neq 0$, since E is non singular. If $n(y) = 0$, we get $n(x+y) = 2\cdot B(x,y) \neq 0$, since char $F \neq 2$ by assumption.

ii) Let x in E be anisotropic. The subspace Fx of E is non singular. Thus Prop. 1.2 implies that $E = Fx \perp (Fx)^{\perp}$. Now induction on the dimension of the space gives the desired result.

An orthogonal sum $(a_1) \perp \dots \perp (a_r)$ will be abbreviated by $(a_1,\dots,a_r)$.

<u>(*) Remark 1.3.a</u> Theorem 1.3 shows that E has an orthogonal basis if char $F \neq 2$. If char $F = 2$, E has an orthogonal basis if and only if E contains an anisotropic vector (cf.[49], Chapter I, § 3.3 or [12], § 6, Theorem 1). The following example shows that this is not always true.

<u>(*) Example 1.4.</u> Let (E,B) be a binary (i.e. two-dimensional) space and assume that E has a basis e,f with $B(e,e) = B(f,f) = 0$ and $B(e,f) = 1$. For an arbitrary $x = \lambda e + \mu f$ in E we have $n(x) =$ $= \lambda^2 n(e) + \mu^2 n(f) = 0$. All vectors in E are isotropic.

Let us use Theorem 1.3 to classify the isomorphism classes of bilinear spaces in the following special case:

Example 1.5. Let F be a field with char $F \neq 2$, such that $F^* = F^{*2}$. If E is a bilinear space over F, we have $E = (a_1,\dots,a_n)$ by Theorem 1.3. Now $F^* = F^{*2}$ implies $(a_i) \cong (1)$ for every one-dimensional space (a_i). Thus $E = (1,\dots,1) = n \times (1)$ *). The isomorphism class of E is therefore uniquely determined by the dimension of E.

The following theorem of Witt is of central importance in the theory of bilinear spaces:

Theorem 1.6. (Witt [64]) Let E_1, E_2 and G be bilinear spaces over F and char $F \neq 2$. If $E_1 \perp G$ is isomorphic to $E_2 \perp G$, then E_1 is isomorphic to E_2.

Proof. Inducting on the dimension of G, we are reduced to show the following: If x,y are anisotropic vectors of a bilinear space (E,B), there is an isometry $\tau: E \to E$, such that $\tau(x) = y$. Since char $F \neq 2$, the vectors x+y and x-y cannot both be isotropic. For any anisotropic vector a of E we denote by τ_a the reflection $z \to z-2B(z,a)\cdot n(a)^{-1}\cdot a$ at the hyperplane $(Fa)^{\perp}$, which is an isometry from E to E. If x-y is anisotropic the isometry τ_{x-y} takes x to y, while if x+y is anisotropic the isometry $\tau_y\tau_{x+y}$ takes x to y.

Example 1.7. Consider the three-dimensional spaces $E := \begin{pmatrix} 0 & 1 \\ 1 & 0 \end{pmatrix} \perp (a)$ and $G := \begin{pmatrix} a & 1 \\ 1 & 0 \end{pmatrix} \perp (a)$ with a in F^*. If $\{x,y,z\}$ is a basis of E and $\{x',y',z'\}$ a basis of E' corresponding to the given matrix representa-

---

*) For every bilinear space E we denote the orthogonal sum of n copies of E by $n \times E$.

tions map x',y',z' to $x+z,y,z-ay$. This gives an isomorphism from G to E as one easily checks. Theorem 1.6 shows that $\begin{pmatrix} a & 1 \\ 1 & 0 \end{pmatrix} \cong \begin{pmatrix} 0 & 1 \\ 1 & 0 \end{pmatrix}$ for any a in F^*, if char $F \neq 2$. But if char $F = 2$, the space $\begin{pmatrix} 0 & 1 \\ 1 & 0 \end{pmatrix}$ cannot be isomorphic to the space $\begin{pmatrix} a & 1 \\ 1 & 0 \end{pmatrix}$ for a in F^*, since $\begin{pmatrix} 0 & 1 \\ 1 & 0 \end{pmatrix}$ contains no anisotropic vector in this case by Example 1.4. Thus the theorem of Witt is not true in characteristic 2.

The binary space $\begin{pmatrix} 0 & 1 \\ 1 & 0 \end{pmatrix} =: H$ is called the <u>hyperbolic plane.</u> An orthogonal sum of copies of H is called a <u>hyperbolic space</u> and an orthogonal sum of the form $\begin{pmatrix} a_1 & 1 \\ 1 & 0 \end{pmatrix} \perp \dots \perp \begin{pmatrix} a_r & 1 \\ 1 & 0 \end{pmatrix}$ is called a <u>metabolic space.</u> In the following H will always denote the hyperbolic plane. Example 1.7 implies, that in char $F \neq 2$ every metabolic space is isomorphic to a hyperbolic space. The metabolic spaces are examples of isotropic spaces. The following fundamental theorem of Witt shows that an arbitrary space decomposes into an anisotropic and a metabolic space.

<u>Theorem 1.8. (Witt [64])</u> Every bilinear space (E,B) has an orthogonal decomposition $E \cong E_o \perp \begin{pmatrix} a_1 & 1 \\ 1 & 0 \end{pmatrix} \perp \dots \perp \begin{pmatrix} a_r & 1 \\ 1 & 0 \end{pmatrix}$ with E_o anisotropic. The isomorphism class of E_o and the number r are uniquely determined by E.

<u>Proof.</u> If E is already anisotropic, we are done. If not, we can find an isotropic vector x in E. Since E is non singular, there is y in E, such that $B(x,y) = 1$. The subspace of E spanned by y and x has a matrix of the form $\begin{pmatrix} n(y) & 1 \\ 1 & 0 \end{pmatrix}$. Thus E splits into the orthogonal sum of $\begin{pmatrix} n(y) & 1 \\ 1 & 0 \end{pmatrix}$ and a subspace E_1. Now induction proves the existence of the decomposition of E. As we remarked above, if char $F \neq 2$, every metabolic space is isomorphic to a hyperbolic one. Thus we get

$E \cong E_o \perp r \times H$ in this case with E_o anisotropic. Let $E \cong F_o \perp s \times H$ b another decomposition of E with F_o anisotropic and, say, $r \leq s$. Witt' Theorem 1.6 shows that we can cancel hyperbolic planes. Thus we arriv at the isomorphy $E_o \cong F_o \perp (s-r) \times H$. But E_o is anisotropic, thus we get $E_o \cong F_o$ and $s = r$. The proof of the uniqueness of the number r and the isomorphism class of E_o in the case char F = 2 is more difficult. Proofs can be found in [34,§ 8.3] and in [50, Theorem 2].

The decomposition of a space E, referred to in Theorem 1.8, is called the <u>Witt-decomposition</u> of E. E_o is the <u>anisotropic part</u> of E and r the <u>index</u> of E. We write r = ind E.

<u>(*) Remark.</u> If char $F \neq 2$, a space E is obviously uniquely determin up to isomorphism by the isomorphism class of the anisotropic part and by the index. If char F = 2, there is the following result (cf. [34, § 8.3] and [50, Theorem 3]). E is uniquely determined up to isomorphism by the dimension, the isomorphism class of the anisotropic part and the additive subgroup of F generated by the norms n(x) of all x in E.

<u>Proposition 1.9.</u> Every maximal totally isotropic subspace of E has dimension ind E. In particular E is metabolic if and only if E contains a totally isotropic subspace S with dim E = 2 dim S.

<u>Proof.</u> Let S be a totally isotropic subspace of E. We choose a subspace S' of E with $E \cong S^{\perp} \oplus S'$. Then it is easily seen that $N := S \oplus S'$ is metabolic. {Notice $S \subset S^{\perp}$, and that the bilinear form of E yields a duality between S and $E/S^{\perp}$.} We have $E = N \perp G$ with $G := N^{\perp}$. We further choose a Witt decomposition $G = P \perp G_o$ of G, whe P denotes a metabolic space and G_o an anisotropic space. Then clearl

$$2 \text{ ind } E = \dim N + \dim P.$$

We have $\dim N = 2 \dim S$ and $\dim P = 2 \dim U$ with U a totally isotropic subspace of P, thus

$$\text{ind } E = \dim(S \perp U).$$

Assume now that S is a maximal totally isotropic subspace. Then we must have $U = 0$ and $\text{ind } E = \dim S$.

q.e.d.

We quote an application of Prop. 1.9, which to some extent motivates the construction of the Witt ring, given in the second part of this chapter.

<u>Corollary 1.10.</u> Let (E,B) be a space over F. The space $(E,B) \perp (E,-B)$ is metabolic.

Indeed, the "diagonal" Δ of $(E,B) \perp (E,-B)$ is a totally isotropic subspace with $\dim \Delta = \dim E$.

Let (E,B) be a space over F. We say that E <u>represents</u> an element a in F, if there is some x in E with $a = n(x)$. {Recall $n(x) := B(x,x)$} Let $n(E)$ denote the set of elements represented by E. We call E <u>universal</u>, if F^* is contained in $n(E)$. For instance the hyperbolic plane H is universal, if char $F \neq 2$, whereas H represents only zero, if char $F = 2$ (Ex. 1.4).

<u>Lemma 1.11.</u> Let E be a space over F, char $F \neq 2$. E is universal if and only if $E \perp (a)$ is isotropic for every a in F^*.

<u>Proof.</u> Let E be universal and a an arbitrary element of F^*. We can find x in E such that $n(x) = -a$. Thus we get $E \cong E_1 \perp (-a)$, hence

$E \perp (a) \cong E_1 \perp (-a) \perp (a)$, which is clearly isotropic. Conversely, let a be arbitrary in F^* and assume that $E \perp (a)$ is isotropic. Let z be a basis vector of the one-dimensional space (a). There is some x in and λ in F, such that $n(x+\lambda z) = 0$. If $\lambda = 0$, x is isotropic, hence E splits off a hyperbolic plane H, which is universal, since char. $F \neq 2$. If $\lambda \neq 0$, we have $n(x) = -\lambda^2 a$, hence $n(-\lambda^{-1}x) = a$. Thus E represents a. Since a was arbitrary, this proves that E is universal.

We use this lemma to classify bilinear spaces in the following special case:

<u>Prop. 1.12.</u> Let F be a field with char $F \neq 2$ and assume that every binary space over F is universal. Every bilinear space over F is uniquely determined up to isomorphism by dimension and determinant.

<u>Proof.</u> Let (E,B) be an arbitrary space over F with dim E =: n. Sinc char $F \neq 2$, the Witt decomposition of E is of the form $E \cong E_o \perp r \times$ with E_o anisotropic. Since by assumption every binary space is unive sal, Lemma 1.11 implies that dim $E_o \leq 2$. Thus if dim E is odd, E_o is necessarily one-dimensional, $E_o = (a)$, and det(E) = $= (-1)^{\frac{n-1}{2}} a \bmod F^{*2}$. If dim E is even and $E_o = 0$, we have det E = $= (-1)^{n/2} \bmod F^{*2}$. If dim E is even and E_o a binary space, E_o is universal, hence represents 1. We thus get $E_o \cong (1) \perp (a)$ with $a \neq -1 \bmod F^{*2}$, since E_o is anisotropic. In this case we have det E $= (-1)^{n/2} (-a) \bmod F^{*2}$. Thus indeed dimension and determinant determine E up to isomorphism.

The assumptions of Prop. 1.12 are true for instance for finite fields (cf. [51; 62:1] or [44; 0.20]).

§ 2 Witt- and Grothendieck rings.

We denote the set of isomorphism classes (E) of non singular bilinear spaces (E,B) over F by S(F). In S(F) we define addition by (E) + (G) := (E ⊥ G). This turns S(F) into a commutative semigroup. Let us define a multiplication in S(F). If (E_1,B_1) and (E_2,B_2) are bilinear spaces, we equip the tensor product $E_1 \otimes_F E_2$ of E_1 and E_2 with the unique symmetric bilinear form B satisfying

$$B(e_1 \otimes e_2, f_1 \otimes f_2) = B_1(e_1,f_1)\, B_2(e_2,f_2)$$

for e_1,f_1 in E_1 and e_2,f_2 in E_2. {Exercise: Show that B exists.} This form B will be denoted by $B_1 \otimes B_2$, and the space $(E_1 \otimes E_2, B_1 \otimes B_2)$ is called the tensor product of (E_1,B_1) and (E_2,B_2). This bilinear space will usually be denoted more briefly by $E_1 \otimes E_2$.

The following properties of the tensor product are easily verified (cf.[12] or [49]). If E_1 and E_2 are non singular, so is $E_1 \otimes E_2$. Furthermore

$$\dim(E_1 \otimes E_2) = \dim E_1 \cdot \dim E_2,$$

$$\det(E_1 \otimes E_2) = \det E_1^{\dim E_2} \cdot \det E_2^{\dim E_1} .$$

For any (E),(G) in S(F) we define (E)·(G) := (E ⊗ G). Thus S(F) becomes a commutative semiring.

Let us give an example for the tensor product of bilinear spaces.

<u>Example 1.13.</u> For any bilinear space (E,B) and a in F^* the product $(a) \otimes (E,B)$ is isomorphic to $(E, a\cdot B)$. An isomorphism is given as follows: Let z be a basis vector of (a), $n(z) = a$. We map the element $\lambda\cdot z \otimes e$ to $\lambda\cdot e$ for all λ in F and e in E.

As a special case we get $(a) \otimes (b) \cong (a\cdot b)$. Thus the group of square classes $Q(F) := F^*/F^{*2}$ is isomorphic to the multiplicative subgroup of one-dimensional spaces, which we denote by $S_1(F)$.

Theorem 1.6 implies that the additive cancellation law holds in $S(F)$ if char $F \neq 2$. Let us call two bilinear spaces E and G <u>strongly equivalent</u>, written as $E \approx G$, if there is a space M such that $E \perp M$ is isomorphic to $F \perp M$. We introduce the set $\overline{S}(F) = S(F)/\approx$ of strong equivalence classes. We equip $\overline{S}(F)$ with the unique addition and multiplication such that the natural projection from $S(F)$ to $\overline{S}(F)$ is a homomorphism. $\overline{S}(F)$ is again a semiring. From Theorem 1.6 it is clear that strong equivalence is the same as isomorphism if char $F \neq 2$. Thus in this case $\overline{S}(F) = S(F)$, and nothing new has been obtained. But if char $F = 2$ we now have cancellation in $\overline{S}(F)$. We denote the strong equivalence class of a space E by $\lfloor E \rfloor$.

Before we go on to construct a ring out of $\overline{S}(F)$, let us take a closer look at the notion of strong equivalence (of interest only if char $F = 2$). As a consequence of Example 1.7 we note that a metabolic space M with Witt-index r is strongly equivalent to the hyperbolic space $r \times H$. This in turn implies that every space is strongly equivalent to a space that has an orthogonal basis. Thus $\overline{S}(F)$ is additively generated by the classes $[(a)]$ of one dimensional spaces (a). We write more briefly $\lfloor a \rfloor$ instead of $\lfloor (a) \rfloor$.

<u>Prop. 1.14.</u> Let E,G be spaces over F. The following are equivalent:

i) E and G are strongly equivalent

ii) There is a metabolic space M, such that $E \perp M \cong G \perp M$

iii) The isomorphism classes of the anisotropic parts and the Witt-indices of E and G are equal.

<u>Proof.</u> i) $\Rightarrow$ ii) This follows easily from Corollary 1.10.
ii) $\Rightarrow$ iii) This is clear from Theorem 1.8.
iii) $\Rightarrow$ i) Let $E \cong E_o \perp M$, $G \cong E_o \perp N$ be the Witt decompositions of E and G with E_o anisotropic and M,N metabolic. Since the dimensions of M and N are equal, we have $M \approx N$. Hence E is strongly equivalent to G.

Let $\hat{W}(F)$ be the set of all formal expressions $[E] - [G]$ with $[E],[G]$ in $\overline{S}(F)$. Two expressions $[E] - [G]$ and $[E'] - [G']$ are defined to be equal if and only if $[E] + [G'] = [E'] + [G]$, i.e. $E \perp G' \approx E' \perp G$. We define addition and multiplication in $\hat{W}(F)$ as follows:

$$([E_1] - [G_1]) + ([E_2] - [G_2]) := [E_1 \perp E_2] - [G_1 \perp G_2],$$

$$([E_1] - [G_1]) \cdot ([E_2] - [G_2]) :=$$
$$= [(E_1 \otimes E_2) \perp (G_1 \otimes G_2)] - [(E_1 \otimes G_2) \perp (G_1 \otimes E_2)].$$

It is easy to check that we obtain in this way well defined compositions on $\hat{W}(F)$ which turn $\hat{W}(F)$ into a commutative ring. We call $\hat{W}(F)$ the <u>Witt-Grothendieck ring</u> of F. We have a natural injective map $[E] \to [E] - [0]$ of $\overline{S}(F)$ into $\hat{W}(F)$, and we regard $\overline{S}(F)$ as a subset of $\hat{W}(F)$, by this injection. Thus we identify the formal difference $[E] - [0]$ with $[E]$, and we now may read a formal difference $[E] - [G]$ as an honest difference in the ring $\hat{W}(F)$. Our ring $\hat{W}(F)$ has the unit element $[1]$. $\hat{W}(F)$ is generated as an additive group by the elements $[a]$ with a in F^*.

Notice that we gain $\hat{W}(F)$ from $\bar{S}(F)$ in the same way as usually the integers $\mathbb{Z}$ are constructed from the natural numbers $\mathbb{N}$.

Let $i: S(F) \to \hat{W}(F)$ denote the additive and multiplicative map $(E) \to [E]$. This map has the following universal property, as is easily verified: For every additive map $\varphi : S(F) \to A$ into an abelian group A there exists a unique homomorphism $\hat{\Phi}$ from the additive group of $\hat{W}(F)$ to A with $\hat{\Phi} \circ i = \varphi$.

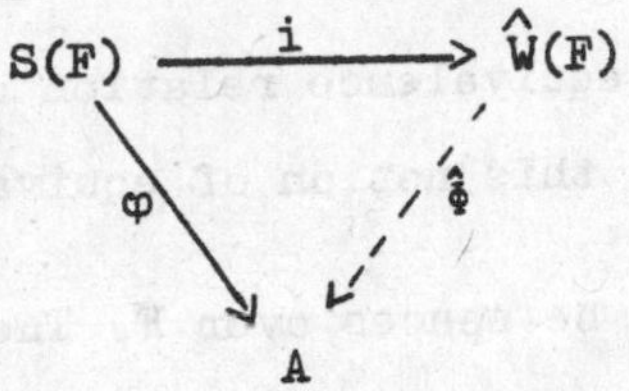

$\{$Define $\hat{\Phi}(\lfloor E\rfloor - \lfloor F\rfloor) = \varphi(E) - \varphi(F).\}$ If A is a ring and φ is also multiplicative, then $\hat{\Phi}$ is a ring homomorphism.

<u>Examples 1.15.</u> We have an additive map $(E) \mapsto \det(E)$ from $S(F)$ to the group of square classes $Q(F)$ of F. Hence we get a group homomorphism from $\hat{W}(F)$ to $Q(F)$, which we again denote by det. In the same way the map $(E) \mapsto \dim(E)$ from $S(F)$ to the natural numbers $\mathbb{N}$ gives a ring homomorphism dim from $\hat{W}(F)$ to $\mathbb{Z}$.

<u>Lemma 1.16.</u> The subgroup $\mathbb{Z} \cdot [H]$ of $\hat{W}(F)$ is an ideal in $\hat{W}(F)$.

<u>Proof.</u> Obviously it is enough to show that $\lfloor a\rfloor \cdot \lfloor H\rfloor = [H]$ for every a in F^*. By example 1.13 we have $(a) \otimes H \cong \begin{pmatrix} 0 & a \\ a & 0 \end{pmatrix}$. But $\begin{pmatrix} 0 & a \\ a & 0 \end{pmatrix}$ clearly is isomorphic to H (exercise).

The quotient $W(F) := \hat{W}(F)/\mathbb{Z}\,\lfloor H\rfloor$ is called the <u>Witt-ring</u> of F. For every space E over F we denote the image of $\lfloor E\rfloor \in \hat{W}(F)$ in $W(F)$ by $\{E\}$. Since by Lemma 1.10 the space $E \perp (-E)$ is metabolic, we have in $W(F)$

$$\{E\} + \{-E\} = 0$$

Thus every element $x = \{E\} - \{F\}$ of W(F) can be represented by a space, $x = \{E \perp (-F)\}$, and not merely by a difference. This fact is the main reason why it is often more convenient to deal with W(F) instead of $\hat{W}(F)$.

<u>Definition.</u> We call two spaces E and G over F <u>equivalent</u>, and write $E \sim G$, if E and G have the same image in W(F), i.e. $\{E\} = \{G\}$.

This is indeed an equivalence relation on S(F) and we have W(F) = = S(F)/~. Let us inspect this notion of equivalence more closely.

<u>Prop. 1.17.</u> Let E and G be spaces over F. The following are equivalent:

i) $E \sim G$.

ii) There is a space W and natural numbers r and s, such that $E \perp W \perp r \times H$ is isomorphic to $G \perp W \perp s \times H$.

iii) There are metabolic spaces M and N, such that $E \perp M$ is isomorphic to $G \perp N$.

iv) The anisotropic parts of E and G are isomorphic.

<u>Proof.</u> i) ⇒ ii) We have $[E] - [G] = s \cdot [H]$, where we may assume $s \geq 0$. Thus E is strongly equivalent to $G \perp s \times H$. This implies ii).
ii) ⇒ iii) Let $E \perp W \perp r \times H$ be isomorphic to $G \perp W \perp s \times H$. Since by lemma 1.10 $W \perp (-W)$ is metabolic, we add (-W) to both spaces and get iii).
iii) ⇒ iv) Evident from Theorem 1.8.
iv) ⇒ i) Clearly $E \sim E_0$ and $G \sim G_0$, where E_0 and G_0 denotes the anisotropic parts of E and G respectively. Since E_0 is isomorphic to G_0, we get $E \sim G$.

As a special case we note that the canonical map from $S_1(F)$ to $W(F)$ is injective.

Part iv) of Prop. 1.17 shows that we could have defined the Witt-ring $W(F)$ as the ring of isomorphism classes of anisotropic spaces. Indeed this was the original definition of Witt (cf.[64]).

Let E and G be two spaces of the same dimension, whose images in $W(F)$ are equal. Then their anisotropic parts are isomorphic. Since E and G have the same dimension, we see moreover that the Witt indices are equal. Hence E and G are strongly equivalent by Prop. 1.14. We get the following corollary:

Corollary 1.18. Let E and G be equivalent spaces of the same dimension. Then E and G are strongly equivalent.

Let $j : S(F) \to W(F)$ denote the canonical map $(E) \mapsto \{E\}$. From the universal property of $\hat{W}(F)$ stated above we obtain the following universal property of this map j. Let $\varphi : S(F) \to A$ be an additive map into an abelian group A with $\varphi(H) = 0$. Then there exists a unique homomorphism Φ from the additive group of $W(F)$ to A with $\Phi \circ j = \varphi$.

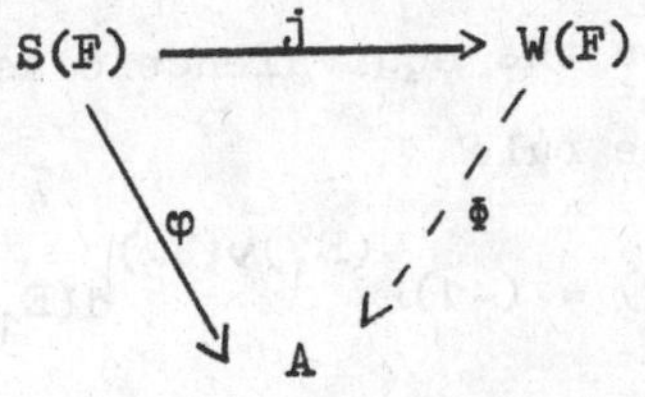

If φ in addition is multiplicative then Φ is a ring homomorphism.

Example 1.19. Unfortunately we have $\dim(H) = 2$ and $\det H = -1$. If we consider the dimension modulo 2, we get a ring homomorphism

$v : W(F) \to \mathbb{Z}/_{2\mathbb{Z}}$, which we call the dimension index. To get something similar to the determinant, we make the following construction. We define on the set $\mathbb{Z}/_{2\mathbb{Z}} \times F^*/_{F^{*2}}$ a multiplication "$\circ$" by $(\mu_1, a_1)\circ(\mu_2, a_2) := (\mu_1 + \mu_2, (-1)^{\mu_1\mu_2} a_1 a_2)$. We thus obtain a group of exponent four, which we denote by $\mathbb{Z}/_{2\mathbb{Z}} \circ F^*/_{F^{*2}}$. The unit element of this group is the pair $(0,1)$. We have a group homomorphism ρ from $\mathbb{Z} \times F^*/_{F^{*2}}$ to $\mathbb{Z}/_{2\mathbb{Z}} \circ F^*/_{F^{*2}}$ defined by $\rho(n,a) := (n \bmod 2, (-1)^{\frac{n(n-1)}{2}} a)$. If we map $S(F)$ to $\mathbb{Z} \times F^*/_{F^{*2}}$ by sending (E) to the pair $(\dim E, \det E)$ and then pass over with ρ to the group $\mathbb{Z}/_{2\mathbb{Z}} \circ F^*/_{F^{*2}}$, we get an additive map from $S(F)$ to $\mathbb{Z}/_{2\mathbb{Z}} \circ F^*/_{F^{*2}}$, which sends (H) to the unit element $(0,1)$. This map is

$$\{E\} \mapsto (\dim E \bmod 2, (-1)^{\frac{n(n-1)}{2}} \det E), \quad n = \dim E.$$

The invariants

$$\nu(E) := \dim E \mod 2$$

and

$$d(E) := (-1)^{\frac{n(n-1)}{2}} \det E$$

are called the <u>dimension index</u> and the <u>signed determinant</u> of the space E. They depend only on the equivalence class $\{E\}$. From our construction we have the rule

$$d(E_1 \perp E_2) = (-1)^{\nu(E_1)\nu(E_2)} d(E_1)d(E_2).$$

Appendix: Quadratic forms

So far we have only considered symmetric bilinear spaces over a field F and never mentioned the notion of a quadratic space over a field. This will be done now. A quadratic space over F is a pair (E,q), where E is a finite-dimensional vector space over F and q is a quadratic form on E, i.e. q is a map from E into F, such that $q(x+y)-q(x)-q(y) =: B_q(x,y)$ is a symmetric bilinear form on $E \times E$ and $q(\lambda x) = \lambda^2 q(x)$ for all λ in F, x in E. The quadratic form q is called non singular, if the associated bilinear form B_q is non singular. If the characteristic of F is not two, we can recover the quadratic form q from the bilinear form B_q, since $q(x) = \frac{1}{2} B_q(x,x)$. Thus in this case the notions of bilinear space and quadratic space actually coincide and we may talk about the Grothendieck- and the Witt-ring of quadratic forms over F. If char F = 2, we have $B_q(x,x) = 2q(x) = 0$. Thus q is not determined by B_q in this case. But nevertheless there exist analoga to the constructions and results above for quadratic forms, as we shall now briefly indicate.

Let F be a field of arbitrary characteristic and denote the set of isomorphism classes of non singular quadratic spaces over F by Sq(F). We have an obvious definition for the orthogonal sum of quadratic spaces. This defines an addition in Sq(F) and we get a semigroup. Arf [3] has shown that in Sq(F) the cancellation law holds (cf. also [12]) in any characteristic. We can construct the "Witt-Grothendieck-group" $\widehat{W}q(F)$ out of Sq(F) in the same way as we did for bilinear spaces. Let $\widetilde{H}$ denote the binary quadratic space Fe + Ef with $q(e) = q(f) = 0$ and $B_q(e,f) = 1$. The quotient group $Wq(F) := \widehat{W}q(F)/\mathbb{Z}\cdot\widetilde{H}$ of $\widehat{W}q(F)$ is called the Witt group of quadratic spaces over F. It is easy to see that for any non singular quadratic space (E,q) the sum (E,q) $\perp$ (E,-q) is hyperbolic, i.e. isomorphic to a direct sum of copies of H. Hence the canonical map from Sq(F) to Wq(F) is surjective. We call two non singular spaces E and G <u>equivalent</u> and write $E \sim G$, if their isomorphism classes have the same image in Wq(F). With this relation we have $Wq(F) = Sq(F)/\sim$. Of course Wq(F) coincides with the additive group of W(F), if char $F \neq 2$. Arf [3] has shown that also for characteristic two there is a Witt decomposition for quadratic spaces. We simply quote this result without proof:

<u>Prop. 1.20.</u> Every quadratic space (E,q) over F has a decomposition $E \cong E_o \perp t \times \tilde{H}$ with E_o anisotropic.

Since cancellation holds true in Sq(F), the isomorphism class of E_o and the number t are uniquely determined by E. E_o is called the <u>anisotropic part</u> of E and t the <u>Witt index</u> of E. A proof similar to that of Prop. 1.17 yields the following proposition:

<u>Prop. 1.21.</u> Two non singular quadratic spaces E and G are equivalent if and only if their anisotropic parts are isomorphic.

We have a natural operation of W(F) on Wq(F) making Wq(F) a W(F)-module: Let (E_1,B_1) be a bilinear space and (E_2,q_2) a quadratic space over F. The tensor product $E_1 \otimes_F E_2$ has a quadratic form q, which is characterized by $B_q = B_1 \otimes B_{q_2}$ and

$$q(e_1 \otimes e_2) := B_1(e_1,e_1)q_2(e_2)$$

for e_1 in E_1 and e_2 in E_2 (cf.[49]; Chap.I, § 8). If B_1 and q_2 are non singular, so is q, and if B_1 or q_2 is hyperbolic, then q is hyperbolic. Thus indeed we obtain a W(F)-module structure on Wq(F).

Chapter II: The structure of Witt rings.

In this chapter we study the ring structure of Witt rings. Starting with a theorem of Witt, which describes the Witt ring of a field F as a quotient of the group ring $\mathbb{Z}[Q(F)]$, where Q(F) denotes the group F^*/F^{*2} of square classes of the field F, we deduce the structure theorems purely ring-theoretically. Thus the results obtained by this way apply to a wider class of rings, which we call "abstract Witt rings", and not only to Witt rings of symmetric bilinear forms over fields (cf. [36]). The main theorems about the structure of Witt rings of fields have been proved by Pfister [52], Leicht-Lorenz [43] and Harrison [32]. Most of the following can be found in a more general setting in [36].

§ 1 Generators and Relations.

In Chapter I we saw that the group Q(F) of square classes of a field F is isomorphic to the group $S_1(F)$ of isomorphism classes of one-dimensional bilinear spaces (cf. Example 1.13). As has been stated in Chap.I the group $S_1(F)$ embeds into the group $W(F)^*$ of units of W(F) and of course it embeds into $\hat{W}(F)^*$ as well. Hence we get ring homomorphisms φ and $\hat{\varphi}$ from the group ring $\mathbb{Z}[Q(F)]$ to W(F) and $\hat{W}(F)$ respectively. As we mentioned in Chapter I, § 2, every space is strongly equivalent to a space with an orthogonal basis. Hence φ and $\hat{\varphi}$ are surjective.

Theorem 2.1. (Witt [64]): The kernel of $\hat{\varphi}$ is additively generated by the elements $(a_1) + (a_2) - (a_1') - (a_2')$ with a_i and a_i' in F^*, $i = 1,2$, and the relation $(a_1) \perp (a_2) \cong (a_1') \perp (a_2')$.

For the proof of this we need the following more geometric theorem. We call two orthogonal bases L and L' of a bilinear space connectable, if there are orthogonal bases $L_1, \dots, L_r$ such that $L_1 = L$, $L_r = L'$ and L_i and L_{i+1} differ at most in two elements for $i = 1, \dots, r-1$.

Theorem 2.2. (Witt [64]): Two orthogonal bases of a bilinear space over F are connectable if F is different from $\mathbb{F}_2$, the field with two elements.

Proof. We restrict ourselves to the case char F $\neq$ 2. A proof of the other case can be found in [34] or [36]. Let $L = \{x_1, \dots, x_n\}$ and $L' = \{y_1, \dots, y_n\}$ be two orthogonal bases of a space E. We show that L is connectable to a basis containing y_1. Then the theorem follows immediately by induction on n. Renumbering the y_i if necessary we assume $y_1 = r_1x_1 + \dots + r_sx_s$ with $s \leq n$ and r_i in F* for $i = 1, \dots, s$. Since clearly L and $\{r_1x_1, \dots, r_sx_s, \dots, x_n\}$ are connectable, we may assume $r_i = 1$ for $i = 1, \dots, s$. We now proceed by induction on s. If $s = 1$ we are done. Let $s \geq 2$. We have $n(y_1) = n(x_1) + \dots + n(x_s) \neq 0$. If we had $n(x_i) + n(x_j) = 0$ for all $i \neq j$, $i,j \leq s$, this would imply $2n(x_i) = 0$ for $i = 1, \dots, s$, hence $n(x_i) = 0$ for $i = 1, \dots, s$, since char F $\neq$ 2. But this is impossible. Thus $n(x_i) + n(x_j) \neq 0$ for some i,j, and we may assume $n(x_1) + n(x_2) \neq 0$. Let $x_1' = x_1 + x_2$ and choose x_2' such that $Fx_1 \perp Fx_2 \cong Fx_1' \perp Fx_2'$. Thus $\{x_1', x_2', x_3, \dots, x_n\}$ is connectable to L, and $y_1 = x_1' + x_3 + \dots + x_s$ is a shorter presentation of y_1.

q.e.d.

Proof of Theorem 2.1. Assume first F $\neq \mathbb{F}_2$ and let x be in ker $\hat{\varphi}$. Then x is of the form $\sum_{i=1}^{r}(a_i) - \sum_{j=1}^{s}(b_j)$ with $(a_1,\dots,a_r)$ strongly equivalent to $(b_1,\dots,b_s)$. Clearly $r = s$ and we can add a space G to $(a_1,\dots,a_r)$ and $(b_1,\dots,b_r)$ such that $(a_1,\dots,a_r) \perp G$ is isomorphic to $(b_1,\dots,b_r) \perp G$. We may assume in addition that G has an orthogonal basis, cf. Th. 1.3 and Remark 1.3.a. Thus we may write $x = \sum_{i=1}^{n}(c_i) - \sum_{j=1}^{n}(d_j)$ with $(c_1,\dots,c_n)$ isomorphic to $(d_1,\dots,d_n)$. There exists a space E over F which has orthogonal bases $L := \{x_1,\dots,x_n\}$ and $L' := \{y_1,\dots,y_n\}$ with $n(x_i) = c_i$ and $n(y_i) = d_i$ for $i = 1,\dots,n$. By Theorem 2.2 the bases L and L' are connectable. Hence there are orthogonal bases $L = L_o, L_1,\dots,L_r = L'$ such that L_i and L_{i+1} differ by at most two elements, $i = 0,\dots,r-1$. Let $L_i = \{z_1^i,\dots,z_n^i\}$ and define

$$(L_i) := \sum_{j=1}^{n}(n(z_j^i)).$$

This is an element of the group ring $\mathbb{Z}[Q(F)]$ and we have

$$x = \sum_{j=0}^{r-1}((L_i)-(L_{i+1})).$$

Each summand $(L_i) - (L_{i+1})$ is of the form $(a_1) + (a_2) - (a_1') - (a_2')$ with $(a_1) \perp (a_2) \cong (a_1') \perp (a_2')$. We finally consider the case $F = \mathbb{F}_2$ We have $F^* = F^{*2}$, hence $\mathbb{Z}[Q(F)] \cong \mathbb{Z}$. The existence of the dimension function on $\hat{W}(\mathbb{F}_2)$ then implies $\hat{W}(\mathbb{F}_2) \cong \mathbb{Z}$, and the kernel of $\hat{\varphi}$ is zero On the other hand any element $(a_1) + (a_2) - (a_1') - (a_2')$ is already zero. Thus Theorem 2.1 is true in this case as well.

<u>Corollary 2.3.</u> i) The kernel of $\hat{\varphi}$ is generated as an ideal in $\mathbb{Z}[Q(F)]$ by the elements $((1)+(a))\ ((1)-(1+a))$ with $a \neq 0,-1$.
ii) $\ker\varphi = \ker\hat{\varphi} + \mathbb{Z}((1)+(-1))$.

Proof. i) It follows from Theorem 2.1 that ker $\hat{\varphi}$ is generated as an ideal by all elements (1) + (b) - (c) - (d) with (1,b) isomorphic to (c,d). Thus $c = \lambda^2 + \mu^2 b$ with λ, μ in F and (b) = (cd), since the determinants must be equal. This implies (d) = (b)·(c). Hence the elements

$$z = ((1)+(b))\ ((1)-(\lambda^2+\mu^2 b))$$

with b in F*, λ,μ in F generate ker $\hat{\varphi}$, the kernel of $\hat{\varphi}$. If $\lambda = 0$, we have z = 0. If $\lambda \neq 0$, the generator z equals ((1)+(b')) ((1)-(1+b')) with $b' = \lambda^{-2}\mu^2 b$.

ii) ker φ is the inverse image in $\mathbb{Z}[Q(F)]$ under $\hat{\varphi}$ of the ideal $\mathbb{Z}\cdot[H]$ of $\hat{W}(F)$. This proves the second statement, since $\hat{\varphi}$ maps (1) + (-1) onto [H].

§ 2 The prime ideals of a Witt ring.

We have seen in the preceding section that we can represent the Witt-Grothendieck- and the Witt ring of bilinear forms over a field F as quotients of the group ring $\mathbb{Z}[G]$, where $G = Q(F)$ is an abelian group of exponent 2. In the following we consider an arbitrary abelian group G of exponent 2 and determine the prime ideal of the group ring $\mathbb{Z}[G]$ and of certain quotients of $\mathbb{Z}[G]$. Any prime ideal in $\mathbb{Z}[G]$ arises as the kernel of a ring homomorphism ψ from $\mathbb{Z}[G]$ into an integral domain T. For every g in G we have $g^2 = 1$, hence $\psi(g) = \pm 1$. Thus ψ maps G always onto the subring $\mathbb{Z}\cdot 1_T$ of T, and we may replace T by $\mathbb{Z}$ or by $\mathbb{Z}/p\mathbb{Z}$ for some prime number p. Now both $\mathbb{Z}$ and $\mathbb{Z}/p\mathbb{Z}$ have no automorphisms, hence different homomorphisms from $\mathbb{Z}[G]$ to any of them will have different prime ideals as kernels. We see that the prime ideals of the group ring $\mathbb{Z}[G]$ are in one-one correspondence with the homomorphisms from $\mathbb{Z}[G]$ to $\mathbb{Z}$ and $\mathbb{Z}/p\mathbb{Z}$. Clearl the ring homomorphisms from $\mathbb{Z}[G]$ to a ring T correspond uniquely with the group homomorphisms from G to T^*. Consider first the case $T = \mathbb{Z}/_{2\mathbb{Z}}$. Since $T^* = \{1\}$ in this case, there is a unique ring homomorphism $\mu : \mathbb{Z}[G] \to \mathbb{Z}/2\mathbb{Z}$, mapping every g in G to 1. We denote the prime ideal $\ker \mu$ by M_o. Clearly M_o contains the element 2, and by the discussion above M_o is the only prime ideal of $\mathbb{Z}[G]$ with this property. We consider now the cases that T equals $\mathbb{Z}$ or $\mathbb{Z}/p\mathbb{Z}$ with $p \neq 2$. Since G has exponent 2 and T^* contains precisely two elements $u = \pm 1_T$ with $u^2 = 1$ in these cases, the homomorphisms from G to T^* correspond uniquely with the complex valued characters of G. We denote a character and its prolongation to a $\mathbb{Z}$-valued homomorphism of $\mathbb{Z}[G]$ by the same letter. We now have the following. Every character χ of G determines a prime ideal $P_\chi := \ker(\chi : \mathbb{Z}[G] \to \mathbb{Z})$, and for every prime $p \neq 2$ it determines a prime ideal $M_{\chi,p} :=$ $\ker(\mathbb{Z}[G] \xrightarrow{\chi} \mathbb{Z} \to \mathbb{Z}/p\mathbb{Z})$. Clearly $M_{\chi,p} = P_\chi + p\cdot\mathbb{Z} = \chi^{-1}(p\mathbb{Z})$. The prime ideals P_χ correspond uniquely to the characters χ and the prime ideals $M_{\chi,p}$ for $p \neq 2$ correspond uniquely to pairs (χ,p) with $p \neq 2$. Since obviously $M_o = P_\chi + 2\mathbb{Z}$ for all characters χ, we use the notations M_o and $M_{\chi,2}$ interchangably in this case. Let us summarize.

<u>Prop. 2.4.</u> Let G be an abelian group of exponent 2. The prime spectrum of the group ring $\mathbb{Z}[G]$ consists of the following prime ideals:

i) $M_o = M_{\chi,2} = \ker(\mu:\mathbb{Z}[G] \to \mathbb{Z}/2\mathbb{Z})$. M_o is the only prime ideal containing 2.

ii) $P_\chi = \ker(\chi:\mathbb{Z}[G] \to \mathbb{Z})$, χ a character of G.

iii) $M_{\chi,p} = \ker(\mathbb{Z}[G] \xrightarrow{\chi} \mathbb{Z} \to \mathbb{Z}/p\mathbb{Z})$, χ a character of G, $p \neq 2$.

$M_{\chi,p}$ is the only prime ideal containing p and P_χ.

Furthermore, different characters χ,ψ yield different prime ideals P_χ, P_ψ, and different pairs (χ,p), (ψ,q) yield different prime ideals $M_{\chi,p}$, $M_{\psi,q}$, provided $p \neq 2$. The P_χ are minimal prime ideals and the $M_{\chi,p}$ are maximal ideals. Every maximal ideal $M_{\chi,p}$ with $p \neq 2$ contains a unique minimal prime ideal, while M_o contains all minimal prime ideals.

Let K be an ideal in the group ring $\mathbb{Z}[G]$, where G is as always an abelian group of exponent 2. The prime ideals of the quotient ring $\mathbb{Z}[G]/K$ correspond one-one to the prime ideals of $\mathbb{Z}[G]$ containing K. In order to decide which prime ideals contain K, we look at the values of the characters of G on K. If, for instance, $\chi(K) = 0$ for a character χ, we have K in P_χ, and if $\chi(K) \subset p\cdot\mathbb{Z}$ we have K in $M_{\chi,p}$.

We shall impose restrictions on the ideals K, we want to consider. Here we are motivated by the following example:

Let G be the group of square classes of a field F. We have $\hat{W}(F) = \mathbb{Z}[G]/\hat{K}$, where $\hat{K}$ is generated by elements $g_1 + g_2 - g_3 - g_4$ with g_i in G for $i = 1,\dots,4$ and with a certain relation. This was proved in Theorem 2.1. In particular we have $g_1 g_2 = g_3 g_4$. It is easily seen that 0 and ± 4 are the only values assumed by a character χ of G on the generators. If we consider the Witt ring W(F) a character χ of G may also assume the value 2 on the generator $\langle 1\rangle + \langle -1\rangle$ of ker φ. From this we conclude:

<u>Prop. 2.5.</u> Let F be a field, $G = Q(F)$ the group of square classes of F and let $\hat{K}$ resp. K denote the kernel of $\hat{\varphi}: \mathbb{Z}[G] \to \hat{W}(F)$ resp. $\varphi: \mathbb{Z}[G] \to W(F)$. We have $\chi(\hat{K}) = 0$ or $4\mathbb{Z}$ and $\chi(K) = 0$, $2\mathbb{Z}$ or $4\mathbb{Z}$ for all characters of G.

<u>Definition 2.6.</u> Let G be an abelian group of exponent 2. A <u>Witt ring for</u> G is a ring $R \neq 0$ together with an isomorphism $\mathbb{Z}[G]/K \xrightarrow{\sim} R$, where the ideal K fulfils the following condition:

$$\chi(K) = 0 \text{ or } = 2^{n(\chi)}\mathbb{Z} \text{ with } n(\chi) \geq 0$$

for every character χ of G. All these rings R are called <u>abstract Witt rings</u>.

We observe that in the condition about the $\chi(K)$ actually the value $n(\chi) = 0$ is impossible. Indeed, since $R \neq 0$ there is a maximal ideal $M_{\chi,p}$ of $\mathbb{Z}[G]$ containing K. Thus $\chi(K)$ is contained in $p\mathbb{Z}$. Since by assumption $\chi(K)$ equals zero or $2^{n(\chi)}\mathbb{Z}$, we get $\chi(K) = 0$ or $p = 2$. In any case K is contained in $M_o = M_{\chi,2}$. Let now χ' be an arbitrary character of G. Since $M_o = M_{\chi',2}$ contains K, we get $\chi'(K) \subset 2\mathbb{Z}$.

We consider the prime spectrum of an abstract Witt ring.

<u>Theorem 2.7.</u> An abstract Witt ring R has precisely the following prime ideals:

i) A unique prime ideal containing $2\cdot 1_R$, which we denote by $M_{o,R}$.
ii) For every $\mathbb{Z}$-valued homomorphism σ of R the prime ideal $P_\sigma := \ker \sigma$. We have $P_\sigma \cap \mathbb{Z} = 0$.
iii) For every $\mathbb{Z}$-valued homomorphism σ of R and every prime p the maximal ideal $M_{\sigma,p} := \sigma^{-1}(p\mathbb{Z})$. $M_{\sigma,p}$ is the only prime ideal containing $p\cdot 1_R$ and P_σ.

$M_{\sigma,2}$ equals $M_{o,R}$ for every σ. Furthermore different homomorphisms σ,τ yield different prime ideals P_σ, P_τ, and different pairs (σ,p), (τ,q) yield different maximal ideals $M_{\sigma,p}$, $M_{\tau,q}$, provided $p \neq 2$. If the set of ring homomorphisms $\mathrm{Hom}(R,\mathbb{Z})$ is non empty, then the P_σ are the minimal prime ideals of R, and the $M_{\sigma,p}$ are the maximal ideals of R. Every $M_{\sigma,p}$ with $p \neq 2$ contains only one minimal prime ideal, namely P_σ, while $M_{o,R}$ contains all minimal prime ideals.

<u>Remark.</u> If $\mathrm{Hom}(R,\mathbb{Z}) = \emptyset$ then by this theorem R is a local ring with a unique prime ideal $M_{o,R}$.

<u>Proof of Th.2.7.</u> By definition up to isomorphy $R = \mathbb{Z}[G]/K$ with G of exponent 2 and $\chi(K) = 0$ or $2^{n(\chi)}\mathbb{Z}$, $n(\chi) \geq 1$, for every character χ of G. As we noted above, the ideal K is contained in M_o. Hence there is a unique prime ideal $M_{o,R} := M_o/K$ in R containing $2\cdot 1_R$. Let Q be an arbitrary prime ideal of R, $Q = P/K$ with $P \supset K$ a prime ideal of $\mathbb{Z}[G]$. By Prop.2.4 there are three cases to consider. i) $P = M_o$. Then Q equals $M_{o,R}$. ii) $P = P_\chi$ for some character χ of G. This implies

$\chi(K) = 0$. Thus χ induces a homomorphism σ from R to $\mathbb{Z}$ and Q equals $P_\sigma := \ker \sigma$. iii) $P = M_{\chi,p}$ for $p \neq 2$. This implies $\chi(K) \subset p\,\mathbb{Z}$. Since $p \neq 2$ and since R is an abstract Witt ring, we again have $\chi(K) = 0$. Thus χ induces a homomorphism σ from R to $\mathbb{Z}$ and Q equals $M_{\sigma,p} := \sigma^{-1}(p\,\mathbb{Z})$. The rest of the statements of the theorem is an easy consequence of Prop.2.4.

<u>Example 2.8.</u> If R is the Witt ring W(F) of a field F, the unique map from R to $\mathbb{Z}/2\,\mathbb{Z}$ is just the dimension index ν considered in Ex. 1.19. Thus the ideal $M_{o,R}$ is just the ideal of all forms of even dimension. We denote $M_{o,R}$ for $R = W(F)$ by I(F) and call it the <u>fundamental ideal</u> of W(F).

A ring R is called a <u>Jacobson-ring</u>, if every prime ideal is an intersection of maximal ideals.

<u>Corollary 2.9.</u> Every abstract Witt ring for a group G is a Jacobson-ring.

<u>Proof.</u> We must show that every minimal prime ideal P_σ is an intersection of maximal ideals. Clearly $P_\sigma \subset \bigcap_{p \text{ prime}} M_{\sigma,p}$. Conversely, if we take x from $\bigcap_{p \text{ prime}} M_{\sigma,p}$, we have $\sigma(x)$ in $p\cdot\mathbb{Z}$ for all primes p, hence $\sigma(x) = 0$. Since $P_\sigma = \ker \sigma$, x is contained P_σ.

<u>Remark.</u> More generally, any homomorphic image of a group ring $\mathbb{Z}[G]$ with G an abelian torsion group is a Jacobson ring. This follows from [13; § 3, Prop5].

As an easy consequence of Corollary 2.9 we see that the nil-radical and the Jacobson radical of an abstract Witt ring coincide.

Let us now consider the situation more closely, if R is the Witt ring W(F) of a field F. We call a homomorphism from W(F) to $\mathbb{Z}$ a signature of F. Let G = Q(F) be the group of square classes of F. We have $W(F) \cong \mathbb{Z}[G]/K$, where K is generated by the elements (1) + (-1) and $((1)+(a))\cdot((1)-(1+a))$ with $a \neq 0,-1$, as we proved in Corollary 2.3. From above we know that any signature of F comes from a character of G, that vanishes on K. But a character χ of G = Q(F) corresponds one-one to a homomorphism σ_χ from F* to $\{\pm 1\}$, and χ vanishes on K, if and only if $\sigma_\chi(-1) = -1$ and $\sigma_\chi(1+a) = 1$ if $\sigma_\chi(a) = +1$ for all a in F*, $a \neq -1$. We see that the signatures of F correspond one-one to the homomorphisms σ from F* to $\{\pm 1\}$ with $\sigma(-1) = -1$ and $\sigma(1+a) = 1$ for every a in F* with $\sigma(a) = 1$. Because of this correspondence we call such homomorphisms from F* to $\{\pm 1\}$ signatures of F as well. We have the following theorem.

Theorem 2.10. (Harrison [32], Leicht-Lorenz [43]): The signatures of a field F are in one-one correspondence to the orderings of F. The correspondence is given by assigning to an ordering "<" of F the signature $\sigma_<$, defined by $\sigma_<(a) = 1$ if $a > 0$ and $\sigma_<(a) = -1$ if $a < 0$.

Proof. Clearly $\sigma_<$ is a signature of F. Let now σ be an arbitrary signature of F. We must define an ordering "<" of F such that $\sigma = \sigma_<$. Let $\mathfrak{P}$ denote the set of all elements a in F* with $\sigma(a) = 1$. Let a be an element of F*, not contained in $\mathfrak{P}$. Thus $\sigma(a) = -1$ and $\sigma(-a) = \sigma(a)\cdot\sigma(-1) = 1$, since $\sigma(-1) = -1$. Hence we see that F is the disjoint union of $\mathfrak{P}$, $-\mathfrak{P}$ and $\{0\}$. Clearly we have $\mathfrak{P}\cdot\mathfrak{P}$ contained in $\mathfrak{P}$. Since $\sigma(1+a) = 1$ if $\sigma(a) = 1$, we have furthermore $1 + \mathfrak{P}$ contained in $\mathfrak{P}$. Thus $\mathfrak{P}$ has all properties of a set of positive elements of an ordering of F, hence defines an ordering < of F such that $a > 0$ if and only if a is in $\mathfrak{P}$ (cf.[14; §§ 1,2]), hence if and only if $\sigma(a) = 1$.

A field F is called real if it has at least one ordering, otherwise F is called non real *). The following corollary is a consequence of the Theorems 2.7 and 2.10.

Corollary 2.11. Let F be a real field. The orderings of F correspond one-one to the minimal prime ideals of W(F).

If F is non real, we know that W(F) is a local ring with I(F) the only prime ideal. The following proposition considers local abstract Witt rings. For any abelian group M we denote by M_t the subgroup of torsion elements of M.

Prop. 2.12. For an abstract Witt ring $R = \mathbb{Z}[G]/K$ the following are equivalent:

i) $M_{o,R}$ is the only prime ideal of R,

ii) $\chi(K) \neq 0$ for all characters χ of G,

iii) $2^n \cdot R = 0$ for some natural number n,

iv) R is torsion, i.e. $R = R_t$,

v) $K \cap \mathbb{Z} \neq \emptyset$.

Proof. The equivalence of i) and ii) is clear from Theorem 2.7.
i) ⇒ iii): $M_{o,R}$ equals the nilradical of R, since it is the only prime ideal. We have $2 \cdot 1_R$ in $M_{o,R}$, hence for some n we have $2^n \cdot 1_R = 0$.
iii) ⇒ iv): trivial, iv) ⇒ v): Since R is torsion, $m \cdot 1_R = 0$ for some natural number m; thus m is in K. v) ⇒ ii): Each character χ of G, interpreted as $\mathbb{Z}$-valued homomorphism of $\mathbb{Z}[G]$, is the identity on $\mathbb{Z}$, hence $\chi(K \cap \mathbb{Z}) = K \cap \mathbb{Z} \neq 0$.

Note, as a special case, that a field F is non real, if and onl if its Witt ring is torsion.

*) In the literature often the term "formally real" is used instead of "real".

§ 3 Nilpotent and torsion elements.

We need the following well-known theorem of Maschke from the representation theory of finite groups.

Theorem 2.13. (Maschke). Let H be a finite group and let F be a field, whose characteristic does not divide the order of H. The F-algebra F[H] is semisimple.

A proof of this theorem can be found in any book on representation theory. Let us consider the group ring $\mathbb{Z}[H]$ of a finite abelian group H and let K be an ideal in $\mathbb{Z}[H]$. We have $\mathbb{Q} \otimes (\mathbb{Z}[H]/K) = \mathbb{Q}[H]/\mathbb{Q} \otimes K$. By Theorem 2.13, the $\mathbb{Q}$-algebra $\mathbb{Q}[H]$ is semisimple, {i.e. $\mathbb{Q}[H]$ is a direct product of finite field extensions of $\mathbb{Q}$}. Hence as a quotient of a semisimple ring, $\mathbb{Q}[H]/\mathbb{Q} \otimes K$ is again semisimple. Thus the nilradical $\mathrm{Nil}(\mathbb{Q}[H]/\mathbb{Q} \otimes K)$ is zero. We use this fact in the following

Lemma 2.14. Let G be an abelian group of exponent 2 and let R = $= \mathbb{Z}[G]/K$ with an arbitrary ideal K of $\mathbb{Z}[G]$. The nilradical Nil(R) is contained in the set R_t of torsion elements of R.

Proof. Let x be an element of Nil(R). We can find a finite subgroup H of G such that x lies in the subring $R_H := \mathbb{Z}[H]/K \cap \mathbb{Z}[H]$ of R. By the preceding remark, $\mathrm{Nil}(\mathbb{Q} \otimes_{\mathbb{Z}} R_H) = 0$. Hence x maps to zero under the canonical map $R_H \to \mathbb{Q} \otimes_{\mathbb{Z}} R_H$, given by $z \mapsto 1 \otimes z$. Since the kernel of this map is precisely $(R_H)_t$, we are done.

Prop. 2.15. Let R be an abstract Witt ring for G and assume $R \neq R_t$. The nilradical Nil R equals R_t.

Proof. In view of Lemma 2.14, we need only to show that R_t is contained in Nil R. Let x be an element of R_t. Hence $n\cdot x = 0$ for some n. Since $R \neq R_t$, there exists a $\mathbb{Z}$-valued homomorphism of R. For any such homomorphism σ we have $n\cdot\sigma(x) = \sigma(nx) = 0$ in $\mathbb{Z}$. Hence $\sigma(x) = 0$, i.e. x is in P_σ. Since Nil R = $\bigcap_\sigma P_\sigma$, we have x in Nil R.

If $R = R_t$, the nilradical equals $M_{o,R}$ by Prop. 2.12. Thus in any case we have Nil R = $R_t \cap M_{o,R}$. We solve now the problem, what kind of torsion may appear in R.

<u>Prop. 2.16.</u> An abstract Witt ring has only 2-torsion.

<u>Proof.</u> If $R = R_t$ we know that $2^nR = 0$ for some natural number n by Prop. 2.12. Thus consider the case $R \neq R_t$ and let x be an element $\neq 0$ with $p\cdot x = 0$ for some prime p. Let σ be any $\mathbb{Z}$-valued homomorphism of R. Since $R \neq R_t$ there is at least one such σ. We have $p\cdot\sigma(x) = \sigma(px) = 0$, thus $\sigma(x) = 0$. Now we may write $x = \sum_{i=1}^{r} a_i$, where a_i equals $\pm \bar{g}_i$ where $\bar{g}_i$ is the image of a group element g_i in R. Since $\sigma(a_i) = \pm 1$ and $\sigma(x) = 0$, the number r must be even, say r = 2n. We may assume that $\sigma(a_i) = 1$ for $1 \leqslant i \leqslant n$ and $\sigma(a_i) = -1$ for $n+1 \leqslant i \leqslant$ Let $y = \prod_{i=1}^{n} (a_i - a_{n+i})$. Then

$$y\cdot x = \prod_{i=1}^{n} (a_i - a_{n+i})\cdot\left(\sum_{i=1}^{n} (a_i + a_{n+i})\right) = 0,$$

since $(a_i - a_{n+i})\cdot(a_i + a_{n+i}) = 1 - 1 = 0$. Thus y lies in the annihilator ann(x) of x and we have $\sigma(y) = 2^n$. We know that ann(x) contains $p\cdot 1_R$. Now ann(x) must be contained in a maximal ideal. By Theorem 2.7 this ideal is necessary of the form M_{p,σ_o} where σ_o is a $\mathbb{Z}$-valued homomorphism of R. Thus we have $\sigma_o(\mathrm{ann}(x)) \subset p\cdot\mathbb{Z}$. Defining the element y above with respect to $\sigma = \sigma_o$ we have $\sigma_o(y) = 2^n$. Thus p = 2.

§ 4 The Theorem of Artin-Pfister.

We use the preceding theorems to attack the problem of finding conditions, under which an element of a field is a sum of squares.

Theorem 2.17. (Artin [4])

Let F be a field, char F $\neq$ 2 and b in F*. The element b is a sum of squares if and only if $\sigma(b) = 1$ for all signatures σ of F, i.e. if $b > 0$ for every ordering ">" of F.
This theorem is a special case of the following.

Theorem 2.18. (Pfister [52])

Let F be a field, char F $\neq$ 2 and let $b, c_1, \ldots, c_r$ be elements in F*. The following are equivalent:

i) $\sigma(b) = 1$ for all signatures σ of F with $\sigma(c_1) = \ldots = \sigma(c_r) = 1$

ii) b is of the form $\sum_{0 \leq i_\nu \leq 1} d_{i_1 \ldots i_r} c_1^{i_1} \ldots c_r^{i_r}$ with all coefficients $d_{i_1}, \ldots, d_{i_r}$ sums of squares.

Proof. The implication ii) $\Rightarrow$ i) is clear. Let G denote the space $\perp_{0 \leq i_\nu \leq 1} (c_1^{i_1} \ldots c_r^{i_r})$. G is just the tensorproduct $(1,c_1) \otimes \ldots \otimes (1,c_r)$, and we must show that b is represented by $N \times G$ for some natural number N. Let E be the space $(1,-b) \otimes G$. For all signatures σ of F we have

$$\sigma(E) = (1-\sigma(b)) \cdot \prod_{i=1}^{r} (1+\sigma(c_i)) = 0$$

since, by assumption, $\sigma(b) = 1$ if $\sigma(c_1) = \ldots = \sigma(c_r) = 1$. Thus the class of E in W(F) is in Nil(W(F)). But Nil(W(F)) is contained in $W(F)_t$ by Lemma 2.14. Thus there is a natural number N, such that $N \times E \sim 0$, i.e. $N \times G \sim (b) \otimes N \times G$. Since both spaces have the same dimension, they are strongly equivalent by Corollary 1.18, hence they are isomorphic, since char $F \neq 2$. The space $N \times G$ represents 1, since G does. The isomorphism $N \times G \cong (b) \otimes (N \times G)$ now implies that b is represented by $N \times G$.

Example 2.19. Let F be a field with char $F \neq 2$. For all signatures σ of F we have $\sigma(-1) = -1$. Thus -1 is a sum of squares if and only if F is non real by the theorem of Artin. Furthermore in this case, any element of F^* is a sum of squares. Notice that in particular all fields of positive characteristic are non real.

We call a field F pythagorean, if any sum of squares is itself a square. Thus, for instance, any field of characteristic 2 and the field of real numbers are pythagorean.

Corollary 2.20. For a field F with char $F \neq 2$ the following are equivalent:

i) F is pythagorean and non real.

ii) $F^* = F^{*2}$.

iii) $W(F) = \mathbb{Z}/2\mathbb{Z}$.

Proof. The equivalence of i) and ii) is clear from Artin's theorem a[nd] the definitions. ii) ⇒ iii): This follows from Ex. 1.5, since the dimension classifies the isomorphy classes of bilinear spaces. iii) ⇒ ii): This is evident since the canonical map from F^*/F^{*2} to $W(F)^*$ is injective.

* Remark. It is easy to see that the equivalence of ii) and iii) above is true in characteristic 2 as well.

We now turn to real pythagorean fields.

Prop. 2.21. For a field F the following are equivalent:

i) F is pythagorean and real.

ii) The Wittring W(F) is torsionfree.

Proof. i) ⇒ ii): It is obviously sufficient to show that the space $N \times E$ is anisotropic, if E is, for all natural numbers N. Let $E = (a_1,\dots,a_r)$. The norm of any element x in $N \times E$ is of the form

$$n(x) = (\sum_{i=1}^{N} y_{i1}^2)a_1 + \dots + (\sum_{i=1}^{N} y_{ir}^2)a_r.$$

Since F is pythagorean we have $\sum_{i=1}^{n} y_{iv}^2 = z_v^2$ for $v = 1,\dots,r$. Now E is anisotropic. Thus $\sum_{v=1}^{r} z_v^2 a_v = 0$ implies that $z_v^2 = \sum_{i=1}^{N} y_{iv}^2 = 0$ for $v = 1,\dots,r$. If we had $y_{iv} \neq 0$ for some i and v, we could write -1 as a sum of squares, a contradiction, since F is real. Thus $n(x) = 0$ implies $x = 0$, i.e. $N \times E$ is anisotropic for any natural number N.

ii) ⇒ i) Clearly $W(F) \neq W(F)_t = 0$. Thus F is real by Prop. 2.12. Let $a = x^2 + y^2$ in F. We must show that a is itself a square. Since a is represented by the space (1,1), we have $(1,1) \cong (a,b)$ for some b in F*. Comparing determinants we obtain (b) = (a). In W(F) we get $2((1) - (a)) = 0$. Since W(F) has no torsion, we conclude (a) = (1), i.e. a is a square.

§ 5 Complements to the structure theory.

In this section we use our main theorems about the prime ideal structure and the torsion of an abstract Witt ring to get results about idempotents, units and zero divisors of an abstract Witt ring and to characterize noetherian Witt rings.

i) Idempotents.

Prop. 2.22. The only idempotents of an abstract Witt ring are 0 and 1.

Proof. Let f be an idempotent of an abstract Witt ring R. Since $f(1-f) = 0$ and $f + (1-f) = 1$, either f or $1 - f$ lies in $M_{o,R}$. Assume f does. Since $1 - f$ is not contained in $M_{o,R}$, it is not contained in any minimal prime ideal of R. Hence f is contained in each minimal prime ideal, hence is nilpotent. This implies $f = 0$.

ii) Units.

Prop. 2.23. Let R be an abstract Witt ring and assume $R \neq R_t$. An element x in R is a unit if and only if $\sigma(x) = \pm 1$ for all $\mathbb{Z}$-valued homomorphisms σ of R.

Proof. Clearly this condition is necessary. Let now $\sigma(x) = \pm 1$ for all $\mathbb{Z}$-valued homomorphisms σ of R. Then we have $\sigma(x^2) = 1$, thus $x^2 - 1$ is in the kernel of all such σ, i.e. $x^2 - 1$ is in Nil(R). Therefore x^2 is of the form $1 + y$ with y in Nil R and hence is a unit (compute the inverse by use of the geometric series). Clearly then x is a unit as well.

If R is a Witt ring for G, we denote the image of an element s of $\mathbb{Z}[G]$ under the pregiven homomorphism from $\mathbb{Z}[G]$ to R by $\overline{s}$. Clearly all elements in the abstract Witt ring $R = \mathbb{Z}[G]/K$ of the form $\pm\, \overline{g}(1+x)$ with g in G and x in Nil(R) are units in R. The following propositions shows, that there are no others.

Prop. 2.24. Let R be a Witt ring for G and let y be a unit in R. Then $y = \pm\, \overline{g}(1+x)$ with g in G and x in Nil(R).

Proof. If R is torsion, we know that R is a local ring with maximal ideal $M_{o,R} = \mathrm{Nil}(R)$. Since $R/M_{o,R} \cong \mathbb{Z}/2\mathbb{Z}$, every unit of R is of the form 1+x with x in Nil(R). Now assume $R \neq R_t$ and let y be a unit in R. We write y in the form $\sum_{i=1}^{m} a_i$ with $a_i = \pm\, \overline{g}_i$ for some g_i in G. We have $\sigma(y) = \pm 1$. Hence $m = 2n + 1$ must be odd and $\sigma(a_i) = 1$ for at least n indices as well as $\sigma(a_i) = -1$ for at least n indices. We thus get $\sigma(y) = \sigma((-1)^n \prod_{i=1}^{m} a_i)$. But $(-1)^n \prod_{i=1}^{m} a_i$ is of the form $\pm\, \overline{g}$ for some g in G. Hence $y = \pm\, \overline{g}z$ with $\sigma(z) = 1$ for all $\mathbb{Z}$-valued homomorphisms σ of R. This implies that $z = 1 + x$ with x nilpotent. Hence $y = \pm\, \overline{g}(1+x)$.

Corollary 2.25. We have $W(F)^* = Q(F)(1+\mathrm{Nil}(W(F)))$.

Proof. This is clear from Prop. 2.19 since Q(F) contains $-1_{W(F)}$.

Remark 2.26. The intersection $Q(F) \cap (1+\mathrm{Nil}(W(F)))$ consists of all square classes (a) with $\sigma(a) = 1$ for all signatures σ of F. By the theorem of Artin (Theorem 2.17) these are all square classes (a) with a a sum of squares.

iii) Zero-divisors

We need the following result from commutative algebra, which goes back to Krull.

Lemma 2.27. Let R be a commutative ring and M an R-module. For the set of zero-divisors N of M in R we have the following:

i) N is a union of prime ideals

ii) If M = R every minimal prime ideal of R is contained in N.

Proof. Let R_S denote the ring of fractions of R with respect to the multiplicative set $S := R \setminus N$. Let a be an element of N. Clearly $Ra \subset N$, hence the ideal $R_S a$ of R_S is different from R_S. Thus there exists a maximal ideal P' of R_S containing $R_S a$. But we have a canonical bijection between the prime ideals of R_S and the prime ideals of R disjoint from S (cf.[16; Chap.II, § 2, Prop.11]). Thus there exists a prime ideal P of R, such that $Ra \subset P \subset N$.

ii) Let P be a minimal prime ideal of R and x in P. Then R_P has the unique prime ideal $P \cdot R_P = \mathrm{Nil}(R_P)$. Thus $\frac{x}{1}$ is nilpotent in R_P. This means that there is an s, not in P, such that $s \cdot x^n = 0$ for some natural number n. Let n be minimal with this property. We have $x \cdot (sx^{n-1}) = 0$. Thus x is in N.

We now return to abstract Witt rings.

Prop. 2.28. All zero-divisors of an abstract Witt ring R lie in $M_{o,R}$.

Proof. By Lemma 2.27 i) the set of zero-divisors of R is a union of prime ideals. Since R has only 2-torsion, the element $p \cdot 1_R$ is not a zero-divisor for $p \neq 2$. But $p \cdot 1_R$ is contained in the maximal ideals $M_{\sigma,p}$ for all $\mathbb{Z}$-valued homomorphisms of R. Thus for $p \neq 2$ the ideal $M_{\sigma,p}$ contains non zero-divisors. Since the remaining prime ideals are all contained in $M_{o,R}$, the set of zero-divisors is contained in $M_{o,R}$.

Remark 2.29. The proof of Prop. 2.28 shows that the set of zero-divisors is equal to $M_{o,R}$ if and only if $R_t \neq 0$, since $2 \cdot 1_R$ lies in $M_{o,R}$, but not in any minimal prime ideal P_σ. If $R_t = 0$ the set of zero-divisors is equal to the union of all minimal prime ideals P_σ by Lemma 2.27 ii).

We now study the Witt rings without zero-divisors.

Prop. 2.30. Let R be an abstract Witt ring without zero-divisors. Then $R \cong \mathbb{Z}$ or $R \cong \mathbb{Z}/2\,\mathbb{Z}$.

Proof. We have an epimorphism $\mathbb{Z}[G] \twoheadrightarrow R$ for some abelian 2-group G, and we denote the image of an element g of G in R by $\overline{g}$. Since

$$(1+\overline{g})(1-\overline{g}) = 0$$

we learn that $\overline{g} = 1$ or $\overline{g} = -1$. Thus R is a homomorphic image of $\mathbb{Z}$. Since R has no odd torsion we must have $R \cong \mathbb{Z}$ or $R \cong \mathbb{Z}/2\,\mathbb{Z}$.

The fields F with $W(F) = \mathbb{Z}/2\,\mathbb{Z}$ had been analyzed in Corollary 2.20. We now take a closer look at the fields F with $W(F) = \mathbb{Z}$.

Prop. 2.31. The following are equivalent for a field F.

i) $W(F) = \mathbb{Z}$.

ii) F is real and has precisely two square classes (1) and (-1).

iii) F has a unique ordering and is pythagorean.

iv) F has a unique ordering, and the positive elements of F with respect to this ordering are the squares $\neq 0$.

Proof. i) $\Rightarrow$ ii): This is clear from the proof of the previous Prop. 2.30.

ii) $\Rightarrow$ i): Clearly $W(F) = \mathbb{Z}(1) \cong \mathbb{Z}$.

(i) $\Rightarrow$ (iii): F is pythagorean, since $W(F) = \mathbb{Z}$ is torsion free. Apparently F has precisely one signature, hence precisely one ordering.

(iii) $\Rightarrow$ (iv): Let $a \in F$ be positive with respect to the unique ordering of R. By the theorem of Artin (2.17) this element a is a sum of square, hence a is itself a square.

The implications (iv) $\Rightarrow$ (iii) and (iii) $\Rightarrow$ (ii) are obvious.

The property (iv) allows to establish in finite dimensional vector spaces over F a euclidean geometry as over the field of real numbers. Thus the fields fulfilling the equivalent properties in Prop. 2.31 are called euclidean.

iv) Noetherian Witt rings

We consider subrings of an abstract Witt ring and want to decide, whether they are noetherian or not.

Lemma 2.32. Let R be an abstract Witt ring, $R \neq R_t$, and let T be a subring of R. For any minimal prime ideal Q of T we have $T/Q \cong \mathbb{Z}$.

Proof. The statement is true for $T = R$. In general, if Q is a minimal prime ideal of T, there exists a minimal prime ideal P of R such that $P \cap T = Q$ (cf.[16], Chap.II, § 2, Prop.16). Thus we get an injection $T/Q \hookrightarrow R/P \cong \mathbb{Z}$. Since $\mathbb{Z}$ has no subrings this is an **isomorphism.**

Prop. 2.33. Let $R = \mathbb{Z}[G]/K$ be an abstract Witt ring. For any subring T of R the following are equivalent:

i) T is noetherian.

ii) There is a finite subgroup H of G such that T is contained in the subring $R_H := \mathbb{Z}[H]/K \cap \mathbb{Z}[H]$ of R.

iii) T is a finitely generated $\mathbb{Z}$-module.

Proof. The equivalence of ii) and iii) and the implication iii) $\Rightarrow$ i) are clear. Thus it suffices to show that a noetherian subring T of an abstract Witt ring R is a finitely generated $\mathbb{Z}$-module. We show first that T/Nil T is a finitely generated $\mathbb{Z}$-module. Since T is noetherian, it has only a finite number of minimal prime ideals $Q_1,\dots,Q_r$ (cf.[16; Chap.II, § 4, Cor.3 of Prop.14]) If $R \neq R_t$ we have $T/Q_i \cong \mathbb{Z}$ by Lemma 2.32. Thus the injection

$$T/\mathrm{Nil}(T) \hookrightarrow \prod_{i=1}^{r} T/Q_i \cong \mathbb{Z} \times \dots \times \mathbb{Z}$$

shows that T/Nil(T) is finitely generated as a $\mathbb{Z}$-module. If $R = R_t$, R has only the prime ideal $M_{o,R}$. Thus Nil $T = M_{o,R} \cap T$ is the only prime ideal of T. We have

$$0 \neq T/\mathrm{Nil}(T) \hookrightarrow R/M_{o,R} \cong \mathbb{Z}/2\mathbb{Z}.$$

Thus $T/\mathrm{Nil}\ T \cong \mathbb{Z}/2\mathbb{Z}$ is again a finitely generated $\mathbb{Z}$-module.

Since T is noetherian, the ideal Nil T is finitely generated. Thus there is a natural number k such that $(\mathrm{Nil}\ T)^k = 0$. We get the following descending chain: $T \supset \mathrm{Nil}\ T \supset (\mathrm{Nil}\ T)^2 \supset \dots \supset (\mathrm{Nil}\ T)^k = 0$.

Since $(\mathrm{Nil}\ T)^i$ is finitely generated for all i, the quotients $(\mathrm{Nil}\ T)^i/(\mathrm{Nil}\ T)^{i+1}$ are finitely generated as T/Nil T-modules for i = 1,...,k-1. But we just proved that T/Nil T is a finitely generated $\mathbb{Z}$-module. Thus all quotients $(\mathrm{Nil}\ T)^i/(\mathrm{Nil}\ T)^{i+1}$ are finitely generated as $\mathbb{Z}$-modules. This implies that T is a finitely generated $\mathbb{Z}$-module.

<u>Prop. 2.34 (Pfister [52]).</u> Let R be the Witt-Grothendieck ring $\hat{W}(F)$ or the Witt ring W(F) of a field F. Then R is noetherian if and only if the group of square classes Q(F) of F is finite.

<u>Proof.</u> If Q(F) is finite, then as already stated above R is noetherian (take T = R). Let now R be noetherian. By Prop. 2.33 there is a finite subgroup H of Q(F) such that R is contained in $R_H = \mathbb{Z}[H]/K \cap \mathbb{Z}[H]$. Without loss of generality we may assume that (-1) is in H. Each g in Q(F) thus has in R_H a presentation $\bar{g} = \sum_{i=1}^{n} \bar{h}_i$ with h_i in H.*) The signed determinant gives us

$$g = (\pm 1) \cdot \prod_{i=1}^{n} h_i .$$ Thus g is already in H. We learn that Q(F) = H is finite.

*) Throughout $\bar{g}$ denotes the image of g in R.

§ 6 Characterization of abstract Witt rings.

Let R be a Witt ring for some abelian group G of exponent 2, and let $\varphi:\mathbb{Z}[G] \twoheadrightarrow R$ denote the associated epimorphism. We call an ideal $\mathfrak{a}$ of R a Witt-ideal, if $R/\mathfrak{a}$ is again a Witt ring for G with respect to $\overline{\varphi}:\mathbb{Z}[G] \twoheadrightarrow R \twoheadrightarrow R/\mathfrak{a}$, the first arrow being φ and the second being the canonical projection from R to $R/\mathfrak{a}$.

Prop. 2.35. An ideal $\mathfrak{a}$ of R is a Witt ideal if and only if $\mathfrak{a} \neq R$ and for every homomorphism σ from R to $\mathbb{Z}$ we have $\sigma(\mathfrak{a}) = 0$ or $\sigma(\mathfrak{a}) = 2^{n(\sigma)}\mathbb{Z}$ with some $n(\sigma) \geq 0$.

Proof. Let K denote the kernal of φ and $\tilde{\mathfrak{a}}$ denote the inverse image $\varphi^{-1}(\mathfrak{a})$. The ring $R/\mathfrak{a}$ is a Witt ring for G if and only if $\mathfrak{a} \neq R$ and $\chi(\tilde{\mathfrak{a}})$ is zero or a power of $2\mathbb{Z}$ for every character χ of G. If $\chi(K) \neq 0$, then $\chi(K)$ is a power of $2\mathbb{Z}$, and thus a priori also $\chi(\tilde{\mathfrak{a}})$ is a power of $2\mathbb{Z}$, since $\tilde{\mathfrak{a}} \supset K$. The proposition now follows from the fact that the characters χ with $\chi(K) = 0$ correspond uniquely with the homomorphisms σ from R to $\mathbb{Z}$.

Remark. If R is a Witt ring for G then in Prop. 2.35 all $n(\sigma)$ must be ≥ 1 (cf. the discussion following Def.2.6).

For an ideal $\mathfrak{a}$ of R we denote by $\sqrt{\mathfrak{a}}$ the radical of $\mathfrak{a}$, i.e. the ideal consisting of all x in R with $x^m \in \mathfrak{a}$ for some $m \geq 1$.

Prop. 2.36. For an ideal $\mathfrak{a}$ of our Witt ring R the following are equivalent:

i) $\mathfrak{a}$ is a Witt ideal for G.

ii) The radical $\sqrt{\mathfrak{a}}$ is a Witt ideal for G.

iii) $\sqrt{\mathfrak{a}} = M_{0,R}$ or $\sqrt{\mathfrak{a}}$ is an intersection of minimal prime ideals of R.

Proof. For every $\mathbb{Z}$-valued homomorphism σ of R we have $\sigma(\mathfrak{a}) \subset \sigma(\sqrt{\mathfrak{a}}) \subset \sqrt{\sigma(\mathfrak{a})}$. Thus $\sigma(\mathfrak{a})$ is zero or a power of $2\mathbb{Z}$ if and only if this holds true for $\sigma(\sqrt{\mathfrak{a}})$. This explains the equivalence (i) $\Leftrightarrow$ (ii). Since now we assume without loss of generality $\mathfrak{a} = \sqrt{\mathfrak{a}}$.
(i) $\Rightarrow$ (iii): Since $\mathfrak{a} = \sqrt{\mathfrak{a}}$, the ideal $\mathfrak{a}$ is an intersection of prime ideals of R. Let $\mathfrak{p}$ be a prime ideal containing $\mathfrak{a}$ which is not minimal. Then by § 2 we have $\mathfrak{p} = M_{\sigma,p}$ for some $\mathbb{Z}$-valued homomorphism of R and prime number p. If $\sigma(\mathfrak{a}) = 0$, then even the minimal prime ideal P_σ contains $\mathfrak{a}$. Otherwise $\sigma(\mathfrak{a})$ is a power of 2 and we must have $p = 2$, hence $\mathfrak{p} = M_{0,R}$. From this observation we learn that $\mathfrak{a}$ is an intersection of minimal prime ideals P_σ or $\mathfrak{a} = M_{0,R}$.
(iii) $\Rightarrow$ (i): Clearly $M_{0,R}$ is a Witt ideal. Let now $\mathfrak{a}$ be an intersection of minimal prime ideals P_{σ_i} with σ_i homomorphisms from R to $\mathbb{Z}$. Let σ be an arbitrary $\mathbb{Z}$-valued homomorphism of R and assume $\sigma(\mathfrak{a}) \neq 0$. We must show that $\sigma(\mathfrak{a}) = 2^{n(\sigma)}\mathbb{Z}$. Obviously it is enough to find an element x in $\mathfrak{a}$ such that $\sigma(x) = 2^n$ for some natural number n. Since $\sigma(\mathfrak{a}) \neq 0$ there is some y in $\mathfrak{a}$ with $\sigma(y) \neq 0$. Write y in the form $y = \sum_{\nu=1}^{n} a_\nu$ with $a_\nu = \pm \bar{g}_\nu$, g_ν elements of the group G. Put $x = \prod_{\nu=1}^{n} (1+\sigma(a_\nu)a_\nu)$. Clearly $\sigma(x) = 2^n$. We want to show that x is in $\mathfrak{a}$. We have

$$\sigma(a_\nu)a_\nu \cdot (1+\sigma(a_\nu)a_\nu) = 1 + \sigma(a_\nu)a_\nu$$

Thus $\sigma(a_\nu)a_\nu \cdot x = x$ or equivalently $a_\nu \cdot x = \sigma(a_\nu)\cdot x$. If we sum over all ν, we get $y \cdot x = (\sum_{\nu=1}^{n} \sigma(a_\nu))\cdot x = \sigma(y)\cdot x$. Now $\sigma(y) \neq 0$. Thus $\sigma(y)$ cannot lie in any minimal prime ideal P_τ, since $P_\tau \cap \mathbb{Z} = 0$. On the other hand $y \cdot x = \sigma(y)x$ is in $\mathfrak{a} = \bigcap_i P_{\sigma_i}$. Thus x is in $\mathfrak{a}$.

Let R be an arbitrary commutative ring. We denote the quotient $R/\text{Nil}\, R$ by R_{red} and call it the <u>reduced ring of R</u>. If R is a quotient $\mathbb{Z}[G]/K$ of a group ring, clearly $R_{red} = \mathbb{Z}[G]/\sqrt{K}$. If we apply the preceding Prop. 2.35 to the case $R = \mathbb{Z}[G]$ and $\mathfrak{a} = K$, we get the following corollary:

<u>Corollary 2.36.</u> A quotient $R = \mathbb{Z}[G]/K$ of a groupring $\mathbb{Z}[G]$ is a Witt ring for G if and only if the reduced ring R_{red} is a Witt ring for G.

We now are able to prove the main result of this section.

<u>Theorem 2.37.</u> Let G be a group of exponent 2. For a quotient $R \neq 0$ of the groupring $\mathbb{Z}[G]$ the following are equivalent:

i) R is a Witt ring for G.

ii) R has only 2-torsion.

<u>Proof.</u> We know by Prop. 2.16 that an abstract Witt ring has only 2-torsion. Now let R be of the form $\mathbb{Z}[G]/K$ and assume that R has only 2-torsion. In view of Prop. 2.35 R is a Witt ring for G if and only if $\sqrt{K}$ is equal to M_o or to an intersection of minimal prime ideals of $\mathbb{Z}[G]$. Thus we must show, that if K is contained in $M_{\chi,p}$ with χ a character of G and p a prime $\neq 2$, then K is already contained in the minimal prime ideal P_χ, or equivalently, if K is contained in $M_{\chi,p}$, $p \neq 2$, then $M_{\chi,p}/K$ is not a minimal prime ideal of R. Since R has only 2-torsion, the element $p \cdot 1_R$ of $M_{\chi,p}/K$ is not a zero-divisor in R for $p \neq 2$. Thus our previous Lemma 2.27 on zero-divisors implies, that indeed $M_{\chi,p}/K$ is not a

minimal prime ideal of R.

We learn from Theorem 2.37 that a commutative ring $R \neq 0$ is an abstract Witt ring if and only if the torsion part R_t is 2-primary and R is a homomorphic image of a group ring $\mathbb{Z}[G]$ for some group G of exponent 2, i.e. if R is generated as a ring over $\mathbb{Z}$ by the elements $x \in R$ with $x^2 = 1$. Then choosing an arbitrary epimorphism $\varphi': \mathbb{Z}[G'] \twoheadrightarrow R$ with G' of exponent 2 we know that R is a Witt ring for G' with respect to φ'.

We close this section with a theorem on the kernel and the cokernel of a homomorphism between abstract Witt rings.

<u>Theorem 2.38.</u> Let $\varphi: R_1 \to R_2$ a homomorphism between commutative rings. Assume that R_1 is an abstract Witt ring, that $R_2 \neq 0$, and that the torsion part $R_{2,t}$ is 2-primary. Then $\varphi(R_1)$ is again an abstract Witt ring. Assume further that R_2 is integral over the subring $\varphi(R_1)$. Then the torsion part of the abelian group $R_2/\varphi(R_1)$ is also 2-primary.

<u>Remark.</u> Clearly all assumptions about R_2 are fulfilled, if R_2 is an abstract Witt ring. Then R_2 is integral even over $\mathbb{Z}$.

<u>Proof of Th. 2.38.</u> The first statement is evident from the previous Theorem 2.37. We denote the image $\varphi(R_1)$ by R and present R as quotient of some group G of exponent 2, $R = \mathbb{Z}[G]/K$. Let x be an element of R_2 and assume $px \in R$ for some prime $p \neq 2$. We have to show that actually x lies in R. By assumption we have an equation

$$x^n + a_{n-1} x^{n-1} + \dots + a_o = 0$$

with coefficients a_i in R. We further can find a finite subgroup H of G such that px is contained in the subring $R_H := \mathbb{Z}[H]/K \cap \mathbb{Z}[H]$ of R.

Since the order of H is a power of 2, R_H/pR_H is a semisimple ring as a consequence of the theorem of Maschke. Hence $\mathrm{Nil}(R_H/pR_H)=0$. Let us multiply the integrality relation for x above by p^n. We get $(px)^n + a_{n-1}\, p(x)^{n-1} + \dots + a_o p^n = 0$. Hence $(px)^n$ is contained in pR_H, i.e. the residue class of px in R_H/pR_H is nilpotent. Since $\mathrm{Nil}(R_H/pR_H) = 0$, the element px actually lies in pR_H. Thus px = py for some y in R_H. This implies $p(x-y) = 0$ in R_2. Since R_2 has only 2-torsion, we get x = y, hence x lies already in R. This finishes the proof of the theorem.

§ 7 Fields with isomorphic Witt rings (cf. [32] and [26]).

We consider homomorphisms between Witt rings over fields and relate them to certain homomorphisms between the square classes of the fields. In this way we attack the problem, to what extent a field is determined by its Witt ring.

In Chapter I, § 2 we have seen that the dimension index ν together with the signed determinant d gives us a group homomorphism (ν, d) from the Witt ring W(F) of a field F to the group $\mathbb{Z}/2\mathbb{Z} \circ Q(F)$. The kernel of ν is just the fundamental ideal I(F) of W(F). Thus we get a homomorphism d' from the abelian group I(F) to Q(F), whose kernel coincides with the kernel of (ν, d). Since $d'(1-(a)) = (a)$ for any a in F*, we see that d' is surjective.

Prop. 2.39 (Pfister [52]). The kernel of d' is equal to $I^2(F)$.*) Thus d' induces an isomorphism $\overline{d}$ from $I(F)/I^2(F)$ onto Q(F).

Proof. I(F) is additively generated by the elements 1 - (a). Indeed, every element of I(F) can be written in the form $-(a_1) - \ldots - (a_{2n})$ with a_i in F* and some $n \geq 1$, and we have for a,b in F*

$$-(a) - (b) = [1-(a)] + [1-(b)] - [1-(-1)]$$

in W(F). Thus $I^2(F)$ is additively generated by the elements $z = [1-(a)][1-(b)]$. For any such generator z we have

$$d(z) = d(1-(a)-(b)+(ab)) = 1.$$

*) We write $I^2(F) = (I(F))^2$.

Thus $I^2(F)$ is contained in the kernel of d'. Let now x be an element of I(F) with d(x) = 1. We have a presentation

$$x = \sum_{i=1}^{r} \varepsilon_i(1-(a_i))$$

with a_i in F* and $\varepsilon_i = \pm 1$. Now

$$-(1-(a)) \equiv 1-(a) \mod I^2(F)$$

and

$$(1-(a)) + (1-(b)) \equiv 1-(ab) \mod I^2(F)$$

for arbitrary a,b in F*. Thus we have

$$x \equiv 1 - (\prod_{i=1}^{r} a_i) \mod I^2(F).$$

From this we obtain

$$d(x) = (\prod_{i=1}^{r} a_i) = 1,$$

and then $x \equiv 0 \mod I^2(F)$. Thus $I^2(F)$ coincides with the kernel of d'.

Remark 2.40. The inverse map of $\overline{d}: I(F)/I^2(F) \xrightarrow{\sim} Q(F)$ clearly is given by $(a) \mapsto 1 - (a) + I^2(F)$.

We now come to the main subject of this section. We consider fields F,L of arbitrary characteristics and a group homomorphism $s: Q(F) \to Q(L)$ between the groups of square classes. Clearly s extends to a ring homomorphism $\tilde{s}$ between the group rings $\mathbb{Z}[Q(F)]$ and $\mathbb{Z}[Q(L)]$. We call s an admissible homomorphism, if $\tilde{s}$ induces a homomorphism from W(F) to W(L), i.e. if $\tilde{s}$ maps the kernel K_F of the canonical projection from $\mathbb{Z}[Q(F)]$ to W(F), described in § 2, into K_L.

Lemma 2.41. For a homomorphism s from Q(F) to Q(L) the following are equivalent:

(i) s is admissible.

(ii) $s(-1) = (-1)$, and for every a in F^* with $s(a) \neq (-1)$ the form $(1,s(a))$ over L represents the square class $s(1+a)$.

Proof. (i) ⇒ (ii): Assume s is admissible. We have

$$\tilde{s}(1+(-1)) = 1 + s(-1) \in K_L,$$

i.e. the form $(1,s(-1))$ over L is equivalent zero. By application of the signed determinant we obtain $s(-1) = (-1)$. Let now (a) be a square class of F with $s(a) \neq (-1)$. Certainly $1 + a \neq 0$. We have

$$\tilde{s}[(1-(1+a))\ (1+(a))] = (1-s(1+a))\ (1+s(a)) \in K_L,$$

hence

$$(1,s(a)) \sim s(1+a) \otimes (1,s(a))$$

over L. Since $s(a) \neq (-1)$ the forms on both side are anisotropic. Thus they are isomorphic, and $s(1+a)$ is represented by $(1,s(a))$.

(ii) ⇒ (i): It follows immediately from the explicit description of K_F in Cor. 2.3 that $\tilde{s}(K_F) \subset K_L$.

We call a homomorphism $s:Q(F) \rightarrow Q(L)$ an admissible isomorphism if s is bijective and both s and s^{-1} are admissible, or equivalently if s induces an isomorphism from W(F) onto W(L).

Theorem 2.42 (Harrison [32]). Let F and L be arbitrary fields. The following are equivalent:

(i) There exists an admissible isomorphism from Q(F) to Q(L).

(ii) The Witt rings W(F) and W(L) are isomorphic.

(iii) $W(F)/I^3(F) \cong W(L)/I^3(L)$.

Proof. The implication (i) ⇒ (ii) is trivial. (ii) ⇒ (iii): Every homomorphisms φ from W(F) to W(L) maps I(F) into I(L). Indeed, $\varphi^{-1}(I(L))$ is a prime ideal of W(F) containing $2 \cdot 1_{W(F)}$, hence $\varphi^{-1}(I(L)) = I(F)$. Thus a homomorphism φ from W(F) to W(L) induces a homomorphism $\overline{\varphi}$ from $W(F)/I^3(F)$ to $W(L)/I^3(L)$. Of course if φ is an isomorphism then the same holds true for $\overline{\varphi}$.

We now enter the proof of (iii) ⇒ (i). The ring $W(F)/I^3(F)$ is an abstract Witt ring, since certainly I(F) is a Witt ideal of W(F). Clearly $I(F)/I^3(F)$ is the prime ideal of $W(F)/I^3(F)$ with residue class field $\mathbb{Z}/2\mathbb{Z}$, and an analogous statement holds true for $W(L)/I^3(L)$. Let ψ be a homomorphism from $W(F)/I^3(F)$ to $W(L)/I^3(L)$. By the same argument as above ψ maps $I(F)/I^3(F)$ into $I(L)/I^3(L)$. Thus ψ also maps $I^2(F)/I^3(F)$ into $I^2(L)/I^3(L)$, hence induces a homomorphism

$$\overline{\psi}: I(F)/I^2(F) \to I(L)/I^2(L).$$

At the beginning of this section we established a canonical isomorphism from Q(F) onto $I(F)/I^2(F)$ mapping a square class (a) of F to $1-(a) + I^2(F)$. Using this isomorphism and the analogous isomorphism from Q(L) to $I(L)/I^2(L)$ we obtain from $\overline{\psi}$ a homomorphism

$$s: Q(F) \to Q(L),$$

which satisfies

$$\text{(A)} \qquad 1-s(a) + I^2(L) = \psi(1-(a)) + I^2(L)$$

(obvious notation). From (A) we want to deduce that s is admissible using property (ii) of Lemma 2.41.

Inserting $a = -1$ in (A) we see that $1 - s(-1)$ lies in $I^2(L)$ and then, applying the signed determinant, that $s(-1) = (-1)$.

Let now a be an element of F^* such that $s(a) \neq (-1)$. With $c := 1 + a$ we have

$$(1-(c))\ (1+(a)) \equiv 0 \mod I^3(F),$$

hence in obvious notation

$$\psi(1-(c))\ \psi(1+(a)) \equiv 0 \mod I^3(L).$$

Since both factors $\psi(1-(c))$, $\psi(1+(a))$ lie in $I(L)$, they both may be altered by summands in $I^2(L)$. Using (A) we get

$$\text{(B)} \qquad (1-s(c))\ (1+s(a)) \in I^3(L).$$

We want to deduce from (B) that

$$\text{(C)} \qquad (1,-s(c)) \otimes (1,s(a)) \sim 0.$$

This is easily done by standard methods of the theory of quadratic forms. Since these methods will not be developed in our lectures, we only indicate the procedure and refer the reader to the literature. If char.L $\neq$ 2 then apply the "Clifford invariant" (cf.[52, § 4], [41, p.120 ff]), to (B) and you see that the quaternion algebra $[s(c), -s(a)]$ over L with structure constants $s(c)$, $-s(a)$ splits.

Thus the norm form $(1,-s(c)) \otimes (1,s(a))$ of this algebra is hyperbolic. In the case char.L = 2 we do not have a Clifford invariant for symmetric bilinear forms, and we resort to the following important theorem of Arason and Pfister (for n = 3):

Theorem 2.43 [1]. Let L be an arbitrary field and φ be an anisotropic form over L whose equivalence class lies in $I^n(L)$, $n \geq 1$. Then $\dim \varphi \geq 2^n$.

By this theorem we obtain from (B) the relation (C) for arbitrary characteristic. Since $s(a) \neq (-1)$ the relation (C) implies

$$(1,s(a)) \cong s(c) \otimes (1,s(a))$$

(cf. proof of Lemma 2.41). Thus s(c) is represented by $(1,s(a))$, and we have proved that s is an admissible homomorphism.

If our homomorphism ψ from $W(F)/I^3(F)$ to $W(L)/I^3(L)$ is an isomorphism, then applying this result also to ψ^{-1} we learn that s is an admissible isomorphism. This finishes the proof of our Theorem 2.42.

Let $s:Q(F) \to Q(L)$ be an admissible homomorphism and assume now that L has characteristic $\neq 2$. Then we can attach to every form φ over F a form $s(\varphi)$ over L in the following way: If char.F = 2 and $\varphi \cong r \times \begin{pmatrix} 0 & 1 \\ 1 & 0 \end{pmatrix}$ for some $r > 1$, then put $s(\varphi) := r \times \begin{pmatrix} 0 & 1 \\ 1 & 0 \end{pmatrix}$. If $\varphi = 0$ then put $s(\varphi) := 0$. In all other cases φ has a diagonalisation

$$\varphi \cong (a_1,\ldots,a_n)$$

and we put

$$s(\varphi) := (s(a_1),\ldots,s(a_n)).$$

This form $s(\varphi)$ does up to isomorphy not depend on the chosen diagonalisation. Indeed, from

$$(a_1,\ldots,a_n) \cong (b_1,\ldots,b_n)$$

we deduce

$$(s(a_1),\ldots,s(a_n)) \sim (s(b_1),\ldots,s(b_n)),$$

since s yields a homomorphism from W(F) to W(L). Since char $L \neq 2$ this implies

$$(s(a_1),\ldots,s(a_n)) \cong (s(b_1),\ldots,s(b_n)).$$

The forms φ and $s(\varphi)$ always have the same dimension.

<u>Prop. 2.44.</u> Assume $s:Q(F) \to Q(L)$ is an admissible homomorphism and char $L \neq 2$. Let φ,ψ,χ be forms over F. Then $\varphi \cong \psi \perp \chi$ implies $s(\varphi) \cong s(\psi) \perp s(\chi)$ and $\varphi \cong \psi \otimes \chi$ implies $s(\varphi) \cong s(\psi) \otimes s(\chi)$.

<u>Proof.</u> We obtain from $\varphi \cong \psi \perp \chi$ that $s(\varphi) \sim s(\psi) \perp s(\chi)$ and then $s(\varphi) \cong s(\psi) \perp s(\chi)$. An analogous argument works for the tensor product.

If φ and ψ are forms over F then we say that φ <u>represents</u> ψ and write $\psi < \varphi$ if $\varphi \cong \psi \perp \chi$ with some other form χ. In other terms $\psi < \varphi$ if and only if there exists an injective homomorphism from the vector space on which ψ lives into the vector space on which φ lives which is compatible with the forms. We say that ψ <u>divides</u> φ and write $\psi | \varphi$ if $\varphi \cong \psi \otimes \chi$ with some other form χ over F.

Obviously our Prop. 2.44 has the following

Corollary 2.45 (cf.[26, § 1]). Assume $s:Q(F) \to Q(L)$ is an admissible homomorphism and char $L \neq 2$. Let φ and ψ be forms over F. Then $\psi < \varphi$ implies $s(\psi) < s(\varphi)$ and $\psi | \varphi$ implies $s(\psi) | s(\varphi)$.

Many properties of quadratic forms studied in the theory over fields up to now can be expressed in such a way, that a combination of non singular forms, built up by orthogonal sums and tensor products, represents or divides some other combination of non singular forms. For example, that a form φ has a non trivial zero means that φ represents $(1,-1)$.

Confining our interest to such properties we obtain from Theorem 2.42, Prop. 2.44 and Cor. 2.45 the following

Meta-theorem (Cordes [26]). Let F and L be fields with characteristic $\neq 2$ which have isomorphic Witt rings. Then the theory of quadratic forms over F is isomorphic to the theory of forms over L.

Indeed, already if $W(F)/I^3(F)$ is isomorphic to $W(L)/I^3(L)$ there exists by Th. 2.42 an admissible isomorphism from Q(F) onto Q(L).

Examples 2.46. Let p be an odd prime number. Let $\mathbb{F}_p((t))$ denote the quotient field of the ring $\mathbb{F}_p[[t]]$ of formal power series in one variable t over the field $\mathbb{F}_p$ consisting of p elements. Further let $\mathbb{Q}_p$ denote the field of p-adic numbers. Then $W(\mathbb{Q}_p)$ and $W(\mathbb{F}_p((t))$ are both known to be isomorphic to the group ring $W(\mathbb{F}_p)[\mathbb{Z}/2\mathbb{Z}]$ (Springer [61]). Thus the forms over $\mathbb{Q}_p$ correspond one-one with the forms over $\mathbb{F}_p((t))$ by an admissible isomorphism. If q is another odd prime with $q \equiv p \mod 4$ then $W(\mathbb{F}_p) \cong W(\mathbb{F}_q)$. Indeed, $W(\mathbb{F}_p)$ is isomorphic to the ring $\mathbb{Z}/4\,\mathbb{Z}$ if $p \equiv 3 \mod 4$ and to the group ring $\mathbb{F}_2[\mathbb{Z}/2\,\mathbb{Z}]$ if $p \equiv 1 \mod 4$, as may be deduced from [51, § 62] (cf. Prop. 1.12). Thus we also have admissible isomorphisms from the group of square classes of $\mathbb{Q}_p$ to the groups of square classes of $\mathbb{Q}_q$ and $\mathbb{F}_q((t))$.

Chapter III Reduced Wittrings

§ 1 Von Neumann regular rings

Let R be an abstract Wittring with $R \neq R_t$. We know by theorem 2.7, that the $\mathbb{Z}$-valued homomorphisms of R correspond one-one to the minimal prime ideals of R. Since $P \cap \mathbb{Z} = 0$ for all minimal prime ideals P of R, the minimal prime ideals correspond one-one to the prime ideals of the localisation of R at $\mathbb{Z} - \{0\}$, and thus to the elements of $\mathrm{Spec}(\mathbb{Q} \otimes_{\mathbb{Z}} R)$. We shall prove that $\mathbb{Q} \otimes_{\mathbb{Z}} R$ is a von Neumann regular ring, i.e. every finitely generated ideal of $\mathbb{Q} \otimes_{\mathbb{Z}} R$ is generated by an idempotent. We use the structure theory of these rings and of their spectra to get more information about abstract Wittrings and their $\mathbb{Z}$-valued homomorphisms. If R is the Wittring of a real field F, the $\mathbb{Z}$-valued homomorphisms of R correspond one-one to the orderings of F. The Zariski-topology on $\mathrm{Spec}(Q \otimes_{\mathbb{Z}} R)$ thus topologizes the set of orderings and we shall determine a subbasis of this topology. In the following we allow us to call a von Neumann regular ring simply "regular", since no confusion can arise. An unadorned "$\otimes$" will always mean a tensor product over $\mathbb{Z}$. A ring is called reduced, if the nilradical is zero.

Prop. 3.1: Let A be commutative ring. The following are equivalent:

i) A is regular.

ii) A is reduced and $\dim A = 0$.

iii) For any a in A there is some x in A with $a^2x = a$.

<u>Proof:</u> i) $\Rightarrow$ ii) Clearly A is reduced. Let P be a prime ideal in A. We must show that P is maximal. Take any q in P and consider the ideal (q,x) generated by q and some x not in P. By assumption (q,x) = (e) with some idempotent e. Clearly e is not in P. Thus 1-e is in P. Now (q,x) is contained in the ideal (P,x) generated by P and x. Thus both e and 1-e are in (P,x), hence (P,x) = A.

ii) $\Rightarrow$ iii) The assumption dim A = 0 implies that the localisation $A_{\mathfrak{m}}$ of A at every maximal ideal $\mathfrak{m}$ has only one prime ideal $\mathfrak{m}A_{\mathfrak{m}}$. Thus $\mathfrak{m}A_{\mathfrak{m}} = \mathrm{Nil}(A_{\mathfrak{m}})$. But since A is reduced, $A_{\mathfrak{m}}$ is reduced. Thus $A_{\mathfrak{m}}$ is a field for every maximal ideal $\mathfrak{m}$. Consider now an arbitrary element a in A. If we localize the principal ideals (a) and (a^2) at a maximal ideal $\mathfrak{m}$ we get 0, if a is contained in $\mathfrak{m}$, and $A_{\mathfrak{m}}$, if a is not contained in $\mathfrak{m}$, since $A_{\mathfrak{m}}$ is a field. Thus the localisations of (a) and (a^2) are equal for all maximal ideals of A. Thus $(a) = (a^2)$, cf. [16; Chap.II, § 3, Th.1], hence there is an element x in A with $a = xa^2$.

iii) $\Rightarrow$ i) By assumption $a = xa^2$ for any a in A. Thus xa is an idempotent, generating the principal ideal (a). If $\mathfrak{a}$ is any finitely generated ideal, $\mathfrak{a}$ is thus generated by a finite number of idempotents. Now observe that the ideal (e,f) generated by two idempotents e and f equals the principal ideal (e+f-ef). Thus $\mathfrak{a}$ is indeed generated by an idempotent.

The characterisation iii) in Prop.3.1 implies at once that any homomorphic image of a regular ring is regular. As a special case we note that every ideal in a regular ring is a radical ideal.

Let us now consider the Zariski-topology on the spectrum Spec A of a regular ring A [16, Chap.II § 4]. As usual we define $V(\mathfrak{a}) = \{P \in \mathrm{Spec}\ A;\ P \supset \mathfrak{a}\}$ for any ideal $\mathfrak{a}$ of A and $D(f) = \{P \in \mathrm{Spec}\ A; f \notin P\}$

for any f in A. It is well-known that the sets D(f) form a basis of the Zariski topology on Spec(A).

Prop. 3.2. Let A be a regular ring.

i) Spec(A) is totally disconnected, i.e. has a basis consisting of sets which are both open and closed. Moreover Spec(A) is compact and Hausdorff.

ii) The sets $V(\mathfrak{a})$ with $\mathfrak{a}$ a finitely generated ideal are open.

iii) We have a bijection $e \to D(e)$ between the idempotents e of A and the clopen (= closed and open) subsets of Spec(A).

A compact totally disconnected Hausdorff space - as our Spec A - will be called in the sequel a Boolean space.

Proof. Let $\mathfrak{a}$ be a finitely generated ideal. Since A is regular, we have $\mathfrak{a} = (e)$ for some idempotent e in A. Thus $V(\mathfrak{a}) = V(e)$ and the relations $e(1-e) = 0$ and $1 = e + (1-e)$ imply $V(e) \cap V(1-e) = \emptyset$ and Spec $A = V(e) \cup V(1-e)$. Thus V(e) is open, hence ii). We see that the sets $D(e) = \mathrm{Spec}(A) \setminus V(e)$ are clopen for all idempotent e of A. Now, if f is any element of A, there is an idempotent e of A such that the ideals (f) and (e) are equal. Thus D(f) equals D(e), and we see that Spec A has a basis of clopen sets. Thus Spec A is totally disconnected. Clearly Spec A is compact, [16, Chap. II, § 4]. Let P and Q be different elements of Spec A. There is an f in P such that f is not in Q. Thus P is in V(f) and Q in D(f). Since V(f) and D(f) are closed and $V(f) \cap D(f) = \emptyset$, we have proved that Spec A is Hausdorff. Let us now prove iii). We assign to each idempotent e of A the clopen set D(e). Let U be an arbitrary clopen set of Spec A. Since the D(e) with e idempotent form a basis of the topology, we may write $U = \bigcup_{i \in I} D(e_i)$. Since Spec A is compact and U closed, U is actually a finite union of $D(e_i)$'s. But the ideal generated by the idempotents e_i, is generated by one idempotent e. Thus U = D(e). It remains to show that our map is injective. Let D(e) be equal to D(f). Thus V(e) = V(f). Since every ideal in A is a radical ideal, we get (e) = (f). Thus for some a in A we have $e = af = af^2 = ef = f$.

<u>Remark:</u> The bijection in Prop. 3.2 iii) remains true for all rings A with dim A = 0, since there are one-to-one correspondences between the prime ideals of A and A_{red} and between the idempotents of A and A_{red}.

So far we have considered arbitrary regular rings. Let R be an abstract Wittring with $R \neq R_t$. Since R is integral over $\mathbb{Z}$, [1] the tensor product $\mathbb{Q} \otimes R$ is an integral $\mathbb{Q}$-algebra. Thus we now characterize regular integral algebras over a field, in order to show that $\mathbb{Q} \otimes R$ is regular.

<u>Prop. 3.3.</u> Let A be an integral algebra over a field F. The following are equivalent:

i) A is regular.

ii) The nilradical of A is zero.

iii) Every finite subalgebra of A is semisimple.

iv) A is an increasing filtered union of semisimple subalgebras,

---

1) i.e. every element of R is integral over $\mathbb{Z}$.

Proof. The implications i) $\Rightarrow$ ii) $\Rightarrow$ iii) are clear. iii) implies iv), since A is integral over F. iv) $\Rightarrow$ i): Let $\mathfrak{a}$ be a finitely generated ideal of A. Thus $\mathfrak{a} = Aa_1 + \ldots + Aa_r$. By assumption there is a semisimple subalgebra B of A such that $a_1,\ldots,a_r$ lie in B. Since B is semisimple, B is a finite product $\prod_{i=1}^{s} F_i$ of fields F_i. Thus the ideal $\mathfrak{b} = Ba_1 + \ldots + Ba_r$ is a subproduct of the fields F_i, hence is generated by an idempotent e. Now $\mathfrak{a} = A\cdot\mathfrak{b} = A\cdot e$, which shows that A is regular.

Let R be an abstract Wittring with $R \neq R_t$. The theorem of Maschke (Th. 2.13) implies that Nil($\mathbb{Q} \otimes R$) is zero. Thus Prop. 3.3 shows that $\mathbb{Q} \otimes R$ is a regular algebra over $\mathbb{Q}$.

Let us change our point of view for a moment. Let X be an arbitrary Boolean space and $A = \mathfrak{C}(X,F)$ the ring of continous functions from X to a discretely topologized field F. We will show that A is a regular algebra over F. Let $\mathfrak{C}$ denote the family of all clopen sets of X. Let χ_U denote the characteristic function of U for U in $\mathfrak{C}$. Clearly the χ_U with U in $\mathfrak{C}$ are all the idempotents of A. Since X is compact, the image of every function f of A in F is compact. Since F is discrete, the image is actually finite. Thus f can be expressed as a finite sum $\sum_{a\in F} a\cdot\chi_{f^{-1}(a)}$. Clearly $f^{-1}(a)$ is in $\mathfrak{C}$. We see that A is generated as an F-module by its idempotents. In particular A is integral over F. Since clearly Nil A = 0, Prop. 3.3 shows that A is a regular F-algebra. Let Y denote the spectrum of A. By Prop. 3.2 i) Y is a Boolean space. Indeed, Y is homeomorphic to the space X, we started with, as the following proposition shows:

Prop. 3.4. Let X be a Boolean space and F a discretely topologized field. The F-algebra $A = \mathfrak{C}(X,F)$ is regular and X is homoeomorphic to Spec A.

Proof: It remains to prove the last assertion. Let Y denote Spec A. We define a map φ from X to Y by assigning to each element x of X the prime ideal P_x, consisting of all functions f in A that vanish in x. Since X is Hausdorff, φ is injective. To show that φ is continuous we only need to verify that the inverse images of the sets D(f) with f in A are open in X. But $\varphi^{-1}(D(f))$ consists of those elements x for which f(x) is not in P_x, i.e. $\varphi^{-1}(D(f))$ is the support of f and hence is open in X. Finally, if D(f) is not empty, the support of f is not empty. Thus the image $\varphi(X)$ of X is dense in Y. Since X is compact and Y is Hausdorff, $\varphi(X)$ is closed in Y, hence $\varphi(X) = Y$, i.e. φ is surjective. Thus φ is a continuous bijective map from the compact space X to the Hausdorff space Y, and hence a homoeomorphism.

The F-algebra $A = \mathfrak{C}(X,F)$ has one property we shall need for our further considerations. If P is a prime ideal in A, P is equal to some P_x, as we have seen in the proof of Prop. 3.4. Thus the evaluation map $f \to f(x)$ from A to the field F defines an isomorphism between A/P and F. The next proposition is a sort of converse to Prop. 3.4:

Prop. 3.5. Let A be a regular F-algebra, such that A/P is isomorphic to F for all prime ideals P of A. Then A is canonically isomorphic to the F-algebra $\mathfrak{C}(X,F)$ where X denotes the prime spectrum of A.

Proof: Since A/P is isomorphic to F for all prime ideals P, we may identify X with the set $Hom_F(A,F)$, letting a homomorphism σ from A to F correspond to the kernel of σ. For every a in A we have a function f_a from X to F defined by $f_a(\sigma) = \sigma(a)$. Let us first show that f_a is continuous. Since F is discretely topologized we only need to verify that $f_a^{-1}(\lambda)$ is open in X for any λ in F. Now $f_a^{-1}(\lambda)$ is the set of all σ in X such that $\sigma(a) = \lambda$, i.e. $\sigma(a-\lambda\cdot 1) = 0$. Thus $f_a^{-1}(\lambda) = V((a-\lambda\cdot 1))$ is open by Prop. 3.2 ii). Thus we obtain a homomorphism ψ from A to $\mathfrak{C}(X,F)$ of F-algebras, defined by $\psi(a) = f_a$. Since $\psi(a) = f_a = 0$ means that a lies in Nil A = 0, ψ is injective. Now let g be an arbitrary element of $\mathfrak{C}(X,F)$. As we noted above, g is a finite linear combination of characteristic functions χ_U with U in $\mathfrak{C}$. According to Prop. 3.2 iii) a set $U \in \mathfrak{C}$ equals D(e) for some idempotent e of A. We have $f_e(\sigma) = 0$ if σ is not contained in D(e) and $f_e(\sigma) = 1$ if σ is contained in D(e). Thus χ_U equals f_e. This shows that ψ is surjective.

The isomorphism ψ from A to $\mathfrak{C}(X,F)$ is often called the <u>Gelfand-isomorphism</u>.

<u>Remarks:</u> i) It is now easy to combine Prop. 3.4 and 3.5 to obtain the following theorem: For any field F the category of Boolean spaces is equivalent to the category of regular F-algebras A with $A/P \cong F$ for all prime ideals P of A. ii) Prop. 3.4 and 3.5 imply that a regular F-algebra A with $A/P \cong F$ for all P in Spec A is already integral over F. If we drop the assumption that Nil A = 0, this conclusion is still true, since A is integral over F if and only if A_{red} is integral over F. For a more general theory see Arens-Kaplansky [2].

As an example let us consider the case $F = \mathbb{Z}/2\,\mathbb{Z}$ more closely. Thus let A be a regular $\mathbb{Z}/2\,\mathbb{Z}$-algebra with $A/P \cong \mathbb{Z}/2\,\mathbb{Z}$ for all P in Spec A. We know that A is generated over $\mathbb{Z}/2\,\mathbb{Z}$ by idempotents. Thus every element of A is a sum of idempotents, hence is itself idempotent, since $2 = 0$ in A. Conversely let A be any commutative ring with $x^2 = x$ for every x in A. Then clearly A is a regular $\mathbb{Z}/2\,\mathbb{Z}$-algebra, and for any prime ideal P of A the quotient A/P is a field which only contains idempotents, thus $A/P \cong \mathbb{Z}/2\,\mathbb{Z}$. Commutative rings with the identity $x^2 = x$ are called "<u>Boolean rings</u>". We have proved:

<u>Corollary 3.6.</u> For a commutative ring A the following are equivalent

i) A is a regular $\mathbb{Z}/2\,\mathbb{Z}$-algebra with $A/P \cong \mathbb{Z}/2\,\mathbb{Z}$ for all P in Spec A.

ii) A is a Boolean ring.

iii) A is isomorphic to $\mathfrak{C}(X, \mathbb{Z}/2\,\mathbb{Z})$ with X a Boolean space.

For any Boolean space X we have a bijection $U \mapsto \chi_U$ from the set $\mathfrak{C}$ to the Boolean ring $\mathfrak{C}(X, \mathbb{Z}/2\,\mathbb{Z})$. Transfering the ring structure from $\mathfrak{C}(X, \mathbb{Z}/2\,\mathbb{Z})$ to $\mathfrak{C}$ by this bijection, we obtain the well-known Boolean addition and multiplication $U + V = (U \cup V) \setminus (U \cap V)$, $U \cdot V = U \cap V$ on $\mathfrak{C}$. In the next section we shall regard $\mathfrak{C}$ as a ring in this way.

§ 2 Topological description of reduced Wittrings.

We now return to abstract Wittrings. Let R be an abstract Wittring with $R \neq R_t$ and let $\bar{R}$ denote the reduced Wittring R_{red}. Since $R/Q \cong \mathbb{Z}$ for all minimal prime ideals Q of R, we have $\mathbb{Q} \otimes R/P \cong \mathbb{Q}$ for all prime ideals P of $\mathbb{Q} \otimes R$. Thus we are in the situation of Prop. 3.5 and $\mathbb{Q} \otimes R$ is isomorphic to $\mathfrak{C}(X,\mathbb{Q})$ with X denoting the Boolean space $\mathrm{Spec}(\mathbb{Q} \otimes R) \cong \mathrm{Hom}(\mathbb{Q} \otimes R,\mathbb{Q}) = \mathrm{Hom}(R, \mathbb{Z}) \cong \mathrm{Hom}(\bar{R}, \mathbb{Z})$. Let T be a subring of R. Lemma 2.32 states that $T/Q \cong \mathbb{Z}$ for all minimal prime ideals Q of T. Thus $\mathbb{Q} \otimes T$ is isomorphic to $\mathfrak{C}(Y,\mathbb{Q})$ with $Y := \mathrm{Spec}(\mathbb{Q} \otimes T)$. There is a canonical surjective map $h: X \to Y$, since any minimal prime ideal Q of T is of the form $P \cap T$ with a minimal prime ideal P of R (cf. [16; Chap. II, § 2, Prop. 16]. The map h induces an injective map h^* from $\mathfrak{C}(Y,\mathbb{Q})$ to $\mathfrak{C}(X,\mathbb{Q})$, assigning to a function f on Y the function $f \circ h$. It is easy to see that h^* is compatible with the Gelfand-isomorphisms, i.e. that the following diagram commutes:

$$\begin{array}{ccc} \mathbb{Q} \otimes T & \xrightarrow{\sim} & \mathfrak{C}(Y,\mathbb{Q}) \\ \big\downarrow & & \big\downarrow h^* \\ \mathbb{Q} \otimes R & \xrightarrow{\sim} & \mathfrak{C}(X,\mathbb{Q}) \end{array}$$

We now consider the reduced Wittring $\bar{R}$. The kernel of the canonical map from R to $\mathbb{Q} \otimes R$ is the torsion group R_t. Since $R \neq R_t$, we have $R_t = \mathrm{Nil}(R)$ by Prop. 2.15. Thus $\bar{R}$ injects into $\mathbb{Q} \otimes R$. Using the Gelfand-isomorphism, we consider $\bar{R}$ as a subring of $\mathfrak{C}(X,\mathbb{Q})$. Thus an element b of $\bar{R}$ corresponds with the function f_b with $f_b(\sigma) = \sigma(b)$

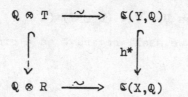

for all σ in $X = \mathrm{Hom}(\mathbb{Q} \otimes R, \mathbb{Q})$. Since $\mathrm{Hom}(\mathbb{Q} \otimes R, \mathbb{Q}) = \mathrm{Hom}(R, \mathbb{Z}) = \mathrm{Hom}(\overline{R}, \mathbb{Z})$, we see that any homomorphism σ from $\mathbb{Q} \otimes R$ to $\mathbb{Q}$ takes values in $\mathbb{Z}$ if we restrict σ to $\overline{R}$. Thus $\overline{R}$ is actually a subring of $\mathfrak{C}(X, \mathbb{Z})$. Now $\mathfrak{C}(X, \mathbb{Z})$ is generated as a $\mathbb{Z}$-module by idempotents. Thus $\mathfrak{C}(X, \mathbb{Z})$ is integral over $\mathbb{Z}$. On the other hand any function f in $\mathfrak{C}(X,\mathbb{Q})$ that is integral over $\mathbb{Z}$ must take values in $\mathbb{Z}$, since $\mathbb{Z}$ is integrally closed. Thus $\mathfrak{C}(X, \mathbb{Z})$ is the integral closure of $\mathbb{Z}$ in $\mathfrak{C}(X,\mathbb{Q})$, hence also the integral closure of $\overline{R}$ in $\mathfrak{C}(X,\mathbb{Q})$. Since $\mathfrak{C}(X, \mathbb{Z})$ and $\overline{R}$ become isomorphic, if we tensor with Q, the group $\mathfrak{C}(X, \mathbb{Z})/\overline{R}$ is torsion. According to Theorem 2.38 this group is 2-primary. Thus we have proved:

<u>Prop. 3.7.</u> Let R be an abstract Wittring with $R \neq R_t$ and X be the Boolean space $\mathrm{Hom}(R, \mathbb{Z}) = \mathrm{Spec}(\mathbb{Q} \otimes R)$.

i) The reduced Wittring $\overline{R}$ is a subring of $\mathfrak{C}(X, \mathbb{Z})$.

ii) $\mathfrak{C}(X, \mathbb{Z})$ is the integral closure of $\overline{R}$ in $\mathbb{Q} \otimes R$.

iii) $\mathfrak{C}(X, \mathbb{Z})/\overline{R}$ is a 2-primary torsion group.

To get more information about $\overline{R}$, we consider at first arbitrary subrings S of $\mathfrak{C}(X, \mathbb{Z})$ for a Boolean space X. In X we have the basis of clopen sets $\mathfrak{C}$. For any U in $\mathfrak{C}$ we define a continuous function φ_U from X to $\mathbb{Z}$ by $\varphi_U = 1 - 2\chi_U$. Thus $\varphi_U(\sigma)$ equals 1 if σ is not contained in U and -1 otherwise. Obviously the φ_U are just the units of $\mathfrak{C}(X, \mathbb{Z})$. We have $\varphi_U \cdot \varphi_V = \varphi_{U+V}$, where + denotes Boolean addition in $\mathfrak{C}$. Thus we get an isomorphism between the group of units $\mathfrak{C}(X, \mathbb{Z})^*$ of $\mathfrak{C}(X, \mathbb{Z})$ and the additive group $\mathfrak{C}$. If S is any subring of $\mathfrak{C}(X, \mathbb{Z})$, clearly $\mathfrak{C}(X, \mathbb{Z})$ is integral over S, since it is integral over $\mathbb{Z}$. Now, if A is a subring of a ring B such that B is integral over A, the units of A are just those units of B, which lie in A. This is an easy consequence of the fact that for any prime ideal P of A there is a prime ideal P' of B with $P = P' \cap A$ (cf. [17; Chap. V, § 2, Th. 1]). Thus in our situation we have $S^* = \mathfrak{C}(X, \mathbb{Z})^* \cap S$, and hence S^* is isomorphic to the subgroups of $\mathfrak{C}$, consisting of all clopen U with φ_U in S^*. We denote this subgroup of $\mathfrak{C}$ by $\mathfrak{D}(S)$. Obviously, $\mathfrak{D}(S)$ may also be described as the set of all U in $\mathfrak{C}$ with $2\chi_U$ in S. Since $\varphi_X(\sigma) = -1$ for all σ in X, the set $\mathfrak{D}(S)$ always contains X. We now characterize the abstract Wittrings lying in $\mathfrak{C}(X, \mathbb{Z})$.

Theorem 3.8. Let R be a subring of $\mathfrak{C}(X, \mathbb{Z})$ for some Boolean space X. The following are equivalent:

i) R is an abstract Wittring.

ii) R is generated as a ring by 1 and all $2\chi_U$ with U in $\mathfrak{D}(R)$.

iii) $R = \mathbb{Z}\cdot 1 + \sum_{U\in\mathfrak{D}(R)} \mathbb{Z}\cdot 2\chi_U$

Furthermore, given any subgroup $\mathfrak{D}$ of $\mathfrak{C}$ with X in $\mathfrak{D}$, there is one and only one abstract Wittring R in $\mathfrak{C}(X, \mathbb{Z})$ such that $\mathfrak{D} = \mathfrak{D}(R)$.

Proof. Every subring R of $\mathfrak{C}(X, \mathbb{Z})$ is torsionfree and every unit has order 2. Thus Prop. 2.37 implies that R is an abstract Wittring if and only if R is generated by its units additively or - what amounts the same - as a ring. Thus the conditions i), ii), iii) are equivalent. To prove the last statement, define $R = \mathbb{Z}\cdot 1 + \sum_{U\in\mathfrak{D}} \mathbb{Z}\cdot 2\chi_U = \sum_{U\in\mathfrak{D}} \mathbb{Z}\,\varphi_U$. Then R is a ring and $\mathfrak{D}$ is contained in $\mathfrak{D}(R)$. Let U be in $\mathfrak{D}(R)$. Thus $2\chi_U$ is in R, hence is of the form $m_o \cdot 1 + \sum_{i=1}^{r} n_i \cdot 2\chi_{U_i}$ with U_i in $\mathfrak{D}$, m_o and n_i integers. Clearly m_o must be an even number $2n_o$, thus $\chi_U = \sum_{i=0}^{r} n_i\chi_{U_i}$ with $U_o = X$. If we consider this equation in $\mathfrak{C}(X, \mathbb{Z}/2\mathbb{Z})$ we obtain $U = \sum_{i=0}^{r} n_iU_i$. Thus U is in $\mathfrak{D}$.

<u>Example 3.9.</u> $\mathbb{Z}\cdot 1 + \mathfrak{C}(X, 2\mathbb{Z})$ is the largest and $\mathbb{Z}\cdot 1$ the smallest abstract Wittring contained in $\mathfrak{C}(X, \mathbb{Z})$.

Let R be an abstract Wittring contained in some $\mathfrak{C}(X, \mathbb{Z})$ for a Boolean space X. We also have the canonical embedding of R in $\mathfrak{C}(Y, \mathbb{Z})$ with $Y := \operatorname{Spec}(\mathbb{Q} \otimes R)$, via the Gelfand isomorphism. There is a continuous map π from $X = \operatorname{Spec}(\mathfrak{C}(X,\mathbb{Q}))$ to Y, which is surjective, since elements of Y correspond to minimal prime ideals of R, which in turn come from minimal prime ideals of $\mathfrak{C}(X,\mathbb{Q})$ (cf. [16; Chap. II, § 2, Prop. 16]). Thus π is identifying and induces an injective map π^* from $\mathfrak{C}(Y,\mathbb{Q})$ to $\mathfrak{C}(X,\mathbb{Q})$. Since π is identifying, we may interpret $\mathfrak{C}(Y,\mathbb{Q})$ as the set of all functions in $\mathfrak{C}(X,\mathbb{Q})$, that are constant on the fibers of π. The diagram

$$\begin{array}{ccc} & & \mathfrak{C}(Y, \mathbb{Z}) \\ & \nearrow & \downarrow \pi^* \\ R & & \\ & \searrow & \\ & & \mathfrak{C}(X, \mathbb{Z}) \end{array}$$

compares the canonical embedding of R in $\mathfrak{C}(Y, \mathbb{Z})$ with the given embedding of R in $\mathfrak{C}(X, \mathbb{Z})$. Obviously π is injective if and only if R separates the points of X. In this case, the map π^* is an isomorphism.

R is determined by the additive subgroup $\mathfrak{D}(R)$ of the Boolean ring $\mathfrak{E}$. What does it mean for $\mathfrak{D}(R)$ that R separates the points of X? The answer will be given below in Corollary 3.12.

It is clear from § 1 that R separates the points of X if and only if $\mathbb{Q}R = \mathfrak{C}(X,\mathbb{Q})$. In general $\mathbb{Q}R$ is only a $\mathbb{Q}$-subalgebra of $\mathfrak{C}(X,\mathbb{Q})$.

Thus we are led to study $\mathbb{Q}$-algebras T of $\mathfrak{C}(X,\mathbb{Q})$ and the sets $\mathfrak{D}(T)$. Slightly more general we replace $\mathbb{Q}$ by some subring Λ of $\mathbb{Q}$ containing $\frac{1}{2}$. (It is sometimes important to work with $\mathbb{Z}[\frac{1}{2}]$ instead of $\mathbb{Q}$.)

<u>Prop. 3.10.</u> Let X be a Boolean space.

i) If T is any Λ-subalgebra of $\mathfrak{C}(X,\Lambda)$ then $\mathfrak{D}(T)$ is a subring of the Boolean ring $\mathfrak{C}$.

ii) If $\mathfrak{D}$ is any subring of $\mathfrak{C}$ then

$$T := \sum_{U\in\mathfrak{D}} \Lambda \chi_U = \sum_{U\in\mathfrak{D}} \Lambda \varphi_U$$

is a Λ-subalgebra of $\mathfrak{C}(X,\Lambda)$ with $\mathfrak{D}(T) = \mathfrak{D}$.

Thus the subrings $\mathfrak{D}$ of $\mathfrak{C}$ correspond one-one with those Λ-subalgebras T of $\mathfrak{C}(X,\Lambda)$ which are generated by characteristic functions χ_U.

The proof of part i) is immediate: Since the function $\frac{1}{2}$ lies in T the group $\mathfrak{D}(T)$ consists of all U in $\mathfrak{C}$ with χ_U in T. Now $\chi_X = 1 \in T$. Moreover if χ_U and χ_V lie in T then also $\chi_{U\cdot V} = \chi_U\chi_V$ lies in T. Thus $\mathfrak{D}(T)$ is a subring of $\mathfrak{C}$.

Part (ii) is obvious from the following

<u>Lemma.</u> Let $\mathfrak{D}$ be a subring of $\mathfrak{C}$ and T the algebra $\sum_{U\in\mathfrak{D}} \Lambda \varphi_U$. Then every f in T can be expressed as a sum $f = \sum_{i=1}^{r} \lambda_i \chi_{U_i}$ with λ_i in Λ and <u>disjoint</u> sets U_i in $\mathfrak{D}$. Thus $f^{-1}(\lambda) \in \mathfrak{D}$ for every λ in Λ.

<u>Proof.</u> Write f in the form $\sum_{k=1}^{t} \mu_k \chi_{V_k}$ with μ_k in Λ and V_k in $\mathfrak{D}$. We have the following partition of X into sets belonging to $\mathfrak{D}$:

$$X = \prod_{i=1}^{t} (V_i + (X+V_i)) = \sum_{r=0}^{t} \sum_{(i_1,\ldots,i_t)} V_{i_1} \ldots V_{i_r} (X+V_{i_{r+1}}) \ldots (X+V_{i_t})$$

with $(i_1,\ldots,i_t)$ running through all permutations of $(1,\ldots,t)$ with $i_k < i_{k+1}$ if $k+1 \leq r$ and if $k > r$. The function f is constant on each set $V_{i_1} \ldots V_{i_r} (X+V_{i_{r+1}}) \ldots (X+V_{i_t})$.

Remark 3.10a. In the case $\Lambda = \mathbb{Q}$ every subalgebra T of $\mathfrak{C}(X,\mathbb{Q})$ is generated by characteristic functions. Indeed, let f be a non zero function in T. Write

$$f = \sum_{i=1}^{r} \lambda_i \chi_{U_i}$$

with $\lambda_i \neq 0$, $\lambda_i \neq \lambda_j$ for $i \neq j$ and disjoint $U_i \in \mathfrak{C}$. We must show that the χ_{U_i} lie in T. By the Chinese remainder theorem applied to the polynomial ring $\mathbb{Q}[t]$ - or by Lagrange's interpolation formula there exist polynomials $p_1(t),\ldots,p_r(t)$ in $\mathbb{Q}[t]$ such that $p_i(\lambda_j) = \delta_{ij}$. We have $\chi_{U_i} = p_i(f) \in T$.

Corollary 3.11. Let R be an abstract Wittring in $\mathfrak{C}(X, \mathbb{Z})$ for some Boolean space X and Λ a subring of $\mathbb{Q}$ containing $\frac{1}{2}$. Then $\mathfrak{D}(\Lambda R)$ is the subring $<\mathfrak{D}(R)>$ of $\mathfrak{C}$ generated by $\mathfrak{D}(R)$.

Proof. ΛR is the smallest Λ-algebra in $\mathfrak{C}(X,\Lambda)$ which contains R and is generated by characteristic functions, and $<\mathfrak{D}(R)>$ is the smallest subring of $\mathfrak{C}$ containing $\mathfrak{D}(R)$. Thus the corollary is a direct consequence of Prop. 3.10.

Corollary 3.12. Let Λ be a subring of $\mathbb{Q}$ containing $\frac{1}{2}$. For an abstrac Wittring R in $\mathfrak{C}(X, \mathbb{Z})$ the following are equivalent:

i) $\mathfrak{C}(X, \mathbb{Z})/R$ is a torsion group.

ii) $\Lambda R = \mathfrak{C}(X,\Lambda)$.

iii) $\mathfrak{H}(R)$ is a subbasis of $\mathfrak{C}$, i.e. $\langle\mathfrak{H}(R)\rangle = \mathfrak{C}$.

iv) R separates the points of X.

Proof. Corollary 3.11 implies the equivalence of ii) and iii). Thus the statement in ii) is independent of the subring Λ of $\mathbb{Q}$ we have chosen. Now ii) with $\Lambda = \mathbb{Q}$ is equivalent to iv) as has been observed above. The equivalence of i) and ii) is clear, since we know a priori that the torsion of $\mathfrak{C}(X, \mathbb{Z})/R$ is 2-primary (Theorem 2.38).

We now come back to the situation we started with at the beginning of this section. Let R be an abstract Wittring with $R \neq R_t$, and consider the reduced Wittring $\overline{R}$ as a subring of $\mathfrak{C}(X, \mathbb{Z})$ with $X := \operatorname{Spec}(\mathbb{Q} \otimes R) = \operatorname{Hom}(R, \mathbb{Z})$. We know by Theorem 3.8 that $\overline{R} = \mathbb{Z} \cdot 1 + \sum_{U \in \mathfrak{H}(\overline{R})} \mathbb{Z} \cdot 2\chi_U$, and, by Corollary 3.12, that $\mathfrak{H}(\overline{R})$ is a subbasis of $\mathfrak{C}$. We study this subbasis more closely:

Prop. 3.13. Let $R = \mathbb{Z}[G]/K$ be an abstract Wittring and $X = \operatorname{Hom}(R, \mathbb{Z})$. $\mathfrak{H}(\overline{R})$ is the system of all sets $W(a) = \{\sigma \in X;\ \sigma(a) = -1\}$ with $a = \pm\overline{g}$, $\overline{g}$ denoting the image of a group element g in R.

Proof. Evidently $\varphi_{W(a)} = f_a = a$. Thus the statement means that every unit of $\overline{R}$ has the form $\pm\overline{g}$ with g in G. But this follows from Prop.2.24 since $\overline{R}$ is reduced.

$\mathfrak{H}(\overline{R})$ is called the Harrison subbasis of X. Indeed, in the case that R is the Wittring of a field F Harrison first proposed to introduce a topology on X using this subbasis (unpublished). In this case we shall write W(a) instead of W((a)) for any a in F*.

We now give a characterization of the reduced Wittrings $\overline{R}$ for which $\mathfrak{H}(\overline{R})$ is equal to $\mathfrak{C}$.

<u>Prop. 3.14.</u> Let $R = \mathbb{Z}[G]/K$ be an abstract Wittring with $R \neq R_t$ and $X = \mathrm{Hom}(R, \mathbb{Z})$. The following are equivalent:

i) $\mathfrak{H}(\overline{R}) = \mathfrak{C}$

ii) $R = \mathbb{Z}\cdot 1 + \mathfrak{C}(X, 2\mathbb{Z})$

iii) (Strong Approximation): For any two disjoint closed subsets Y_1, Y_2 of X, there is an element $a = \pm\overline{g}$ with g in G, such that $\sigma(a) = -1$ for all σ in Y_1 and $\sigma(a) = +1$ for all σ in Y_2.

<u>Proof.</u> The equivalence of i) and ii) follows from Theorem 3.8. i) $\Rightarrow$ iii): Since Y_1 and Y_2 are compact, there is a clopen set U such that Y_1 is contained in U and $Y_2 \cap U = \emptyset$. Since U is in $\mathfrak{C} = \mathfrak{H}(\overline{R})$ Prop. 3.13 shows that U equals W(a) for some $a = \pm\overline{g}$. This element a has the desired property. iii) $\Rightarrow$ i) Let U be an arbitrary clopen set. By assumption there is an $a = \pm\overline{g}$ with g in G such that $\sigma(a) = -$ for all σ in U and $\sigma(a) = +1$ for all σ in $X\setminus U$. Thus $U = W(a)$, hence U is in $\mathfrak{H}(\overline{R})$.

We close this section with a study of the set of values of a finite system of signatures $(\sigma_1,\ldots,\sigma_n)$. We call an abstract Wittring $R = \mathbb{Z}[G]/K$ <u>small for G</u>, if there exists a g in G such that $\overline{g} = -$ in R. For example, if R is the Wittring W(F) of a field F, R is small for F^*/F^{*2}. Note also that any Wittring R is small for the group G(R), consisting of all elements x in R with $x^2 = 1$. If R is small for G, Prop. 3.13 implies that $\mathfrak{H}(\overline{R})$ is the system of all sets $W(\overline{g})$ with g running through G.

Prop. 3.15. Let R be a small Wittring for G and let $\sigma_1, \dots, \sigma_n$ be different $\mathbb{Z}$-valued homomorphisms of R. Let B denote the subring of $\mathbb{Z}^n$ of all $(b_1, \dots, b_n)$ with $b_i \equiv b_j \pmod 2$, i.e. $B = \mathbb{Z} + (2\mathbb{Z})^n$. Then the image of $(\sigma_1, \dots, \sigma_n)$ is contained in B. The following are equivalent:

i) $\operatorname{Im}(\sigma_1, \dots, \sigma_n) = B$.

ii) For any i, $1 \leq i \leq n$, there exists g_i in G with $\sigma_i(\overline{g_i}) = -1$ and $\sigma_j(\overline{g_i}) = +1$ for $j \neq i$.

iii) The characters $\chi_i : G \to \{\pm 1\}$, defined by $\chi_i(g) = \sigma_i(\overline{g})$ are linearly independent in the $\mathbb{F}_2$-vectorspace $\hat{G}$ of all characters on G.

Proof. Let us replace R by $R/\bigcap_{i=1}^{n} \ker \sigma_i$, which is again an abstract Wittring for G, according to Prop. 2.36 and which is reduced. Thus we may assume without loss of generality that $X = \operatorname{Hom}(R, \mathbb{Z}) = \{\sigma_1, \dots, \sigma_n\}$ is finite and R reduced. Now the first assertion follows from the description of R as a subring of $\mathfrak{C}(X, \mathbb{Z})$, given in Theorem 3.8. The equivalence of i) and ii) is easily deduced from the equivalence of ii) and iii) in Prop. 3.14, using the fact that R is small and X is finite. Let us prove ii) $\Rightarrow$ iii). If $\chi_1^{i_1} \cdots \chi_n^{i_n} = 1$ is a relation in $\hat{G}$ with $0 \leq i_\nu \leq 1$ we must show $i_\nu = 0$ for all ν. But applying the equation to the element g_k yields $(-1)^{i_k} = 1$, thus $i_k = 0$. Since k is arbitrary, the characters χ_i are proved to be linearly independent. iii) $\Rightarrow$ ii) Let H be a finite subgroup of G, such that the restrictions $\chi_i|H$ of the characters χ_i to H are still linearly independent in $\hat{H}$. Since H is finite, we have $\hat{\hat{H}} = H$. Thus we can find elements g_i, $1 \leq i \leq n$, with $\chi_i(g_i) = -1$ and $\chi_j(g_i) = +1$ for all $j \neq i$. Clearly this proves ii).

Examples 3.16. 1) Let F be a real field and R = W(F) the Wittring of F. If $\sigma_1,\dots,\sigma_n$ are $\mathbb{Z}$-valued homomorphisms of R corresponding to archimedian orderings on F, then each σ_i yields an order isomorphism of F into the field of real numbers (cf. [14 ; Ex. 11a, p.57]) and thus the "weak approximation theorem" (cf. [5 ; Th. 8, p.10]) shows that condition ii) of Prop. 3.15 is fulfilled. Hence $\mathrm{Im}(\sigma_1,\dots,\sigma_n) = B$.

2) Let R be an arbitrary abstract Wittring with $R \neq R_t$ and let $\sigma_1,\dots,\sigma_n$ be $\mathbb{Z}$-valued homomorphisms of R with $n \leq 3$. R is small for G(R) and condition iii) of Prop. 3.15 is fulfilled. This is clear for $n = 1,2$, since no character χ_i, can be the identity character, and if $n = 3$ a relation $\chi_1\chi_2\chi_3 = 1$ is impossible, since $\chi_i(-1) = -1$. Thus we have $\mathrm{Im}(\sigma_1,\dots,\sigma_n) = B$ for $n \leq 3$.

3) Let F be the field $\mathbb{R}((X))((Y))$ of iterated formal power series in two variables over the reals. By a result of Springer [61] the Wittring W(F) is isomorphic to $\mathbb{Z}[G]$, with G the Klein four group. Hence there are exactly four $\mathbb{Z}$-valued homomorphisms $\sigma_1,\dots,\sigma_4$ of W(F) corresponding to the four characters of G. Since these are linearly dependent, we have $\mathrm{Im}(\sigma_1,\dots,\sigma_4) \subsetneq B$.

As a final result in this section we give a lower bound on $\mathrm{Im}(\sigma_1,\dots,\sigma_n)$:

Prop. 3.17. Let R be an abstract Wittring with $R \neq R_t$. Let $\sigma_1,\dots,\sigma_n$ be distinct $\mathbb{Z}$-valued homomorphisms on R. Let $k = [\frac{n}{2}]$ denote the greatest natural number below $\frac{n}{2}$. We have

$$(2^k\mathbb{Z})^n \subset M_{o,R}/\bigcap_{i=1}^{n} \ker \sigma_i \subset (2\mathbb{Z})^n.$$

Proof. We need only prove that $(2^k \mathbb{Z})^n$ is contained in $M_{o,R}/\bigcap_{i=1}^{n} \ker \sigma_i$. Thus we must find an element x_1 in $M_{o,R}$ such that $(\sigma_1(x_1),\ldots,\sigma_n(x_1)) = (2^k,0,\ldots,0)$. This will suffice, since we can replace σ_1 by any of the σ_i to get elements x_i which image $(0,\ldots,0,2^k,0,\ldots,0)$. To find x_1 let us use induction on n. The case $n \leq 3$ has been established in Example 3.16. Since $[\frac{n-2}{2}] = k-1$ we can find an element y such that $(\sigma_1(y),\ldots,\sigma_n(y)) =$ $= (2^{k-1},0,\ldots,0,*,*)$ by the induction hypothesis, where the asterisks stand for some integers. Choosing $n = 3$ we find an element z with $(\sigma_1(z),\ldots,\sigma_n(z)) = (2,*,\ldots*,0,0)$. Now $x_1 = y \cdot z$ has the desired property.

§ 3 A Nullstellensatz for Witt ideals and a generalization of the theorem of Artin-Pfister.

We give two applications of the results established in the preceding section.

Let $\mathfrak{a}$ be a Witt ideal in an abstract Wittring R (see section 6 of Chap. II) and let X denote Hom(R, $\mathbb{Z}$). To $\mathfrak{a}$ we associate the closed set $V(\mathfrak{a})$ in X. Clearly $V(\mathfrak{a}) \supset V(\sqrt{\mathfrak{a}})$. Let us define a map in the other direction. For any non empty subset Y of X define I(Y) to be the set of all x in $M_{o,R}$ with $\sigma(x) = 0$ for all σ in Y. Clearly I(Y) is an ideal and $I(Y) = I(\overline{Y}) = \sqrt{I(Y)}$. If Y is empty we define $I(Y) = M_{o,R}$.

Prop. 3.18. ("Nullstellensatz") Let R be an abstract Wittring and X = Hom(R, $\mathbb{Z}$).

i) For all Witt ideals $\mathfrak{a}$ in R we have $IV(\mathfrak{a}) = \sqrt{\mathfrak{a}}$

ii) For all subsets Y of X we have $VI(Y) = \overline{Y}$.

In particular the Witt ideals $\mathfrak{a}$ with $\mathfrak{a} = \sqrt{\mathfrak{a}}$ correspond one-one to the closed subsets of X.

Proof. Without loss of generality we may assume that $\mathfrak{a} = \sqrt{\mathfrak{a}}$ and $Y = \overline{Y}$.

i) Clearly $\mathfrak{a}$ is contained in $IV(\mathfrak{a})$. On the other hand $IV(\mathfrak{a})$ is the intersection of those minimal prime ideals P_σ which contain $\mathfrak{a}$, hence is equal to $\mathfrak{a}$ by Prop. 2.36 iii.

ii) Clearly Y is contained in VI(Y). Let σ be an element of X which is not in Y. We must show that σ is not in VI(Y), i.e. we must find an element x in $M_{o,R}$ such that $\sigma(x) \neq 0$, but $\tau(x) = 0$ for all τ in Y

Since Y is closed, there is a characteristic function χ_U with U in $\mathfrak{C}$ such that $\chi_U(\sigma) = 1$, but $U \cap Y = \emptyset$. By Prop. 3.7 iii there is a natural number $n \neq 0$ such that $n \cdot \chi_U$ is contained in the reduced ideal $\overline{M}_{o,R}$. Lifting this element back to R, we get an x with the desired property.

As a second application we generalize theorem 2.18.

<u>Theorem 3.19.</u> Let F be a field with char$F \neq 2$, g be a non zero element of F and M a subset of F^*. The following statements are equivalent:

i) $\sigma(g) = 1$ for all signatures σ with $\sigma(M) = 1$

ii) g is contained in the subsemiring of F generated by M and the set of squares F^2.

<u>Proof.</u> If M is finite, this is just Theorem 2.18. In general, we need only show that i) implies ii). For any subset N of F^* let S(N) denote the set of all signatures σ with $\sigma(N) = 1$. According to the definition of W(a) in Prop. 3.13, for any a in F^* the set $S(\{a\}) = S(a)$ equals $W(-a)$, and hence is clopen. Thus $S(M) = \bigcap_{a \in M} S(a)$ is closed as well. By assumption we have $S(M) \subset S(g)$. Write M as the union of all finite subsets M_α of M. Then we have $S(M) = \bigcap_\alpha S(M_\alpha)$, and we must prove that $S(M_\alpha) \subset S(g)$ for some α, since then we can apply Theorem 2.19 to the finite set M_α. Consider the sets $A_\alpha = S(M_\alpha) \setminus S(g)$. These are all closed, since S(g) is open. We have $\bigcap_\alpha A_\alpha = S(M) \setminus S(g) = \emptyset$. Thus $\{A_\alpha\}$ is a system of closed subsets of Sign(F) with empty intersection. Since Sign(F) is compact, there must be a finite collection of sets A_α with empty intersection. Let M' be

the union of the corresponding sets M_α. Then M' is finite and $S(M') \smallsetminus S(g) = \emptyset$, hence $S(M') \subset S(g)$. Thus g is contained in the semiring generated by F^2 and M', and a fortiori contained in the semiring generated by F^2 and M.

§ 4 When are Wittrings group rings?

We prove the following theorem [63].

Theorem 3.20. Let F be a field with -1 not a square in F, and let G denote the group of square classes Q(F). The following are equivalent:

i) There exists a group H of exponent 2 with $W(F) \cong \mathbb{Z}[H]$.

ii) For every subgroup H of index 2 in G not containing the square class (-1) the natural map from $\mathbb{Z}[H]$ to W(F) is an isomorphism.

iii) The kernel of the natural map from $\mathbb{Z}[G]$ to W(F) is the ideal generated by (1) + (-1).

iv) Every character $\chi: G \to \{\pm 1\}$ with $\chi(-1) = -1$ is a signature of F.

v) For every a in F* with -a not a square class we have

$$F^{*2} + aF^{*2} = F^{*2} \cup aF^{*2},$$

in other terms, every anisotropic binary form over F represents at most two square classes.

Proof [63]

(i) ⇒ (v): Let $f: W(F) \to \mathbb{Z}[H]$ be an isomorphism. For every a in F* we have $(a)^2 = 1$ in W(F), hence $f((a))^2 = 1$. Thus $f((a)) = \pm h$ with some h in H (Cor. 2.25). Now assume that $b = \lambda^2 + a\mu^2$ with some λ, μ in F*. Then in W(F)

$$(1+(a))\,(1-(b)) = 0,$$

hence in $\mathbb{Z}[H]$

$$1 = -f((a)) + f((b)) + f((ab)).$$

Thus either $f((a)) = -1$ or $f((b)) = 1$ or $f((ab)) = 1$. Since f is an isomorphism $f((a)) = -1$ would imply $(a) = (-1)$ contrary to assumption. If $f((b)) = 1$ then $(b) = (1)$, i.e. $b \in F^{*2}$. If $f((ab)) =$ then $(ab) = (1)$, i.e. $b \in aF^{*2}$.

v) ⇒ iii): This follows from our description of the kernel K of the natural map from $\mathbb{Z}[G]$ to W(F) in II, § 1.

iii) ⇒ ii): Clearly the principal ideal K' generated by $(1) + (-1)$ in $\mathbb{Z}[G]$ has intersection zero with $\mathbb{Z}[H]$ for every subgroup H of G not containing (-1). If H has index 2 in G then even $\mathbb{Z}[G] = K' \oplus \mathbb{Z}[H]$.

ii) ⇒ i) is trivial. Thus the equivalence of (i), (ii), (iii), (v) is proved. (iii) ⇒ (iv) is trivial. We finally prove (iv) ⇒ (iii).

Thus assume (iv) holds true. The principal ideal K' generated by $(1) + (-1)$ in $\mathbb{Z}[G]$ is contained in the kernel K of the map from $\mathbb{Z}[G]$ to W(F). On the other hand $\mathbb{Z}[G]/K'$ is isomorphic to the group ring $\mathbb{Z}[G/\{(1),(-1)\}]$ and hence has no nilpotent elements. Thus K' is a radical Witt ideal, hence the intersection of the minimal prime ideals P of $\mathbb{Z}[G]$ which contain K'. These prime ideals P are the kernels P_χ of the characters $\chi: \mathbb{Z}[G] \to \mathbb{Z}$ with $\chi((-1)) = -1$. By assumption all these P_χ contain the ideal K, hence $K = K'$.

The fields with the equivalent properties (i) - (v) are clearly real pythagorean. Diller and Dress proved in [29] among other things the following

<u>Theorem.</u> Let F be a real pythagorean field, and a an element of F^* such that $-a$ is not a square. Then $F^{*2} + aF^{*2}$ coincides with

$F^{*2} \cup (aF^{*2})$ if and only if the field $F(\sqrt{a})$ is again pythagorean.

Thus the fields F with W(F) isomorphic to some group ring $\mathbb{Z}[H]$ are precisely the real pythagorean fields for which all real quadratic extensions are again pythagorean. We call these fields following L.Bröcker [18] "strictly pythagorean". (Ware [63], Brown [21] and Elman-Lam [31] use the term "superpythagorean"). The valuation theory and the Galois theory of the strictly pythagorean fields are now fairly well understood, see [8], [18] and [21].

§ 5 Fields with strong approximation for orderings (cf. [31], [55]).

Let F be a real field and X the space of signatures (i.e. orderings) of F. We say F has the "strong approximation property for orderings", - abbreviated "SAP" -, if for any disjoint closed sets A and B in X there exists an element a of F^* with $\sigma(a) = -1$ for all σ in A and $\sigma(a) = +1$ for all σ in B. As stated in § 2 (Prop. 3.14) these are precisely the fields F for which the image $\overline{W}(F)$ of W(F) in $\mathfrak{C}(X, \mathbb{Z})$ has the maximal possible size $\mathbb{Z} + \mathfrak{C}(X, 2\mathbb{Z})$.

First examples of SAP fields are the algebraic number fields, i.e. finite algebraic extensions of the field $\mathbb{Q}$, as far as they are real. Here our property SAP is a consequence of the weak approximation theorem for absolut values in algebraic number theory [17, VI, § 7.3]. It is then rather easily seen that more generally every real algebraic extension F of $\mathbb{Q}$ is an SAP field. (Notice that F is the union of the algebraic number fields contained in F and X is the projective limit of the spaces of orderings of these number fields.) Another prominent class of SAP fields are the real field extensions of transcendency degree 1 of the field $\mathbb{R}$ of real numbers, cf. [53, Th. 9.4].

We now prove the remarkable fact that a real field F is already SAP if a considerably weaker approximation property can be verified.

Theorem 3.21. (Elman and Lam [31]) Assume that for every clopen subset A of X and every point $\sigma \in X$ not contained in A there exists an element a of F* with $\sigma(a) = -1$ and $\tau(a) = +1$ for all τ in A. Then F has SAP.

Proof [55]. According to Prop. 3.14 we have to show that for every clopen set Y there exists some a in F* such that Y coincides with $W(a) = \{\sigma \in X | \sigma(a) = -1\}$. We first discuss the special case

$$Y = W(a_1) \cap \ldots \cap W(a_n)$$

with some elements $a_1,\ldots,a_n$ of F*. For every point σ in Y there exists according to our assumption an element b of F* with $\sigma(b) = -1$ and $\tau(b) = +1$ for all τ in $X \setminus Y$. This means $\sigma \in W(b) \subset Y$. Since Y is compact we conclude that there exist finitely many elements $b_1,\ldots,b_k$ of F* such that

$$Y = W(b_1) \cup \ldots \cup W(b_k).$$

By repeating some of the a_i or b_j if necessary we may assume $k = n$. Consider the bilinear forms

$$\varphi = \langle 1,-a_1\rangle \otimes \ldots \otimes \langle 1,-a_n\rangle$$

and

$$\psi = \langle 1,b_1\rangle \otimes \ldots \otimes \langle 1,b_n\rangle.$$

If σ is in Y we have $\sigma(\varphi) = 2^n$, $\sigma(\psi) = 0$. If σ is in $X \setminus Y$ we have $\sigma(\varphi) = 0$, $\sigma(\psi) = 2^n$. Thus $\varphi \perp \psi$ and $2^n \times \langle 1\rangle$ have the same signature for all σ in X, and we conclude from II, § 3 that for some $m \geq 0$ we have

$$2^m \times \varphi \perp 2^m \times \psi \sim 2^{m+n} \times \langle 1\rangle,$$

hence

$$2^m \times \varphi \perp 2^m \times \psi \cong 2^{m+n} \times \langle 1\rangle \perp 2^{m+n-1} \times H.$$

Thus the form $2^m \times \varphi \perp 2^m \times \psi$ is isotropic, and there certainly exists an element c of F^* such that $-c$ is represented by $2^m \times \varphi$ and c is represented by $2^m \times \psi$. For every σ in $X \setminus Y$ we have $\sigma(2^m \times \psi) = 2^{m+n}$. This implies $\sigma(c) = +1$, since c appears as one of the coefficients of a suitable diagonalization of $2^m \times \varphi$. For every σ in Y we have $\sigma(2^m \times \varphi) = 2^{m+n}$ and hence for the same reason $\sigma(c) = -1$. Thus $Y = W(c)$.

We now have proved that the Harrison subbasis of X, consisting of the sets W(a), actually is a basis of X. Let Y be an arbitrary clopen subset of X. Then there exist finitely many elements $a_1,\dots,a_n$ of F^* with

$$X \setminus Y = W(a_1) \cup \dots \cup W(a_n)$$

since $X \setminus Y$ is open and compact. Thus

$$Y = W(-a_1) \cap \dots \cap W(-a_n),$$

and, as just shown, $Y = W(c)$ for some c in F^*. This finishes the proof of Theorem 3.21.

Using deeper methods one can even prove that F has SAP if for any four different signatures $\sigma_1,\sigma_2,\sigma_3,\sigma_4$ there exists some a in F^* with $\sigma_1(a) = -1$ and $\sigma_2(a) = \sigma_3(a) = \sigma_4(a) = 1$ [18, Satz 3.2

Let φ be a symmetric bilinear form over F and σ a signature of F. We call φ <u>positive definite</u> (negative definite) at σ if $\sigma(\varphi) = \dim \varphi$ (resp. $\sigma(\varphi) = -\dim \varphi$). Choosing some diagonalization $\varphi \cong \langle a_1,\dots,a_n\rangle$ this means that all a_i are positive with respect to the ordering of F corresponding to σ. If φ is neither positive nor negative definite at σ, i.e. $|\sigma(\varphi)| < \dim \varphi$, we call φ <u>indefinite</u> at σ.

If φ is isotropic then of course φ is "totally indefinite", i.e. indefinite at all signatures of F. Let us call a form φ <u>weakly isotropic</u> if some multiple $m \times \varphi$ is isotropic. Clearly also every weakly isotropic form is totally indefinite.

<u>Definition.</u> We say F has the "Hasse-Minkowski-Property for orderings" - abbreviated HMP - if conversely every totally indefinite form φ over F is weakly isotropic.

<u>Remark.</u> The Artin-Schreier theory ([6],[35],[9]) associates with every signature σ of F an up to isomorphism unique euclidean algebraic extension F_σ, the "real closure" of F at σ, such that σ coincides with the natural map $W(F) \to W(F_\sigma) = \mathbb{Z}$. That φ is indefinite at σ means that φ becomes isotropic over F_σ, or equivalently, that φ becomes weakly isotropic over F_σ. Thus HMP means that a form φ over F which becomes (weakly) isotropic over all real closures is itself weakly isotropic. This property is analogous to the Hasse-Minkowski theorem for number fields [51, Th.66:1] (but notice that the Hasse-Minkowski theorem deals with isotropy instead of weak isotropy). Of course for F a number field HMP holds true and is an immediate consequence of the Hasse-Minkowski theorem.

<u>Theorem 3.22.</u> A real field F has HMP if and only if F has SAP.

<u>Proof</u> [55].

SAP $\Rightarrow$ HMP: Let φ be a form over F of dimension n which is totally indefinite. The possible values $\sigma(\varphi)$ for some σ in X are

$-n+2, -n+4, \ldots, n-2$. We define the clopen sets $(1 \leq k \leq n-1)$

$$Y_k := \{\sigma \in X \mid \sigma(\varphi) = -n+2k\}.$$

They are disjoint and cover the whole of X. Since F has SAP there exist elements $b_1, \ldots, b_{n-2}$ in F* such that

$$W(b_k) = Y_1 \cup \ldots \cup Y_k. \quad (1 \leq k \leq n-2)$$

Consider the form $\psi := \langle b_1, \ldots, b_{n-2}\rangle$. Let σ be an element of Y_k with $1 \leq k \leq n-2$. Since Y_k is contained in $W(b_k), \ldots, W(b_{n-2})$ but disjoint to $W(b_1), \ldots, W(b_{k-1})$ we have

$$\sigma(\psi) = (k-1) - (n-2-k+1) = -n+2k.$$

For σ in Y_{n-1} we have

$$\sigma(\psi) = n-2 = -n+2(n-1).$$

Thus $\sigma(\psi) = \sigma(\varphi)$ for all σ in X and we conclude using II, § 3 that $2^m \times \varphi \sim 2^m \times \psi$ for some $m \geq 0$. Since $\dim \psi < \dim \varphi$ the form $2^m \times \varphi$ is isotropic.

HMP ⇒ SAP: We prove that for any two elements a,b of F* there exists some c in F* with $W(a) \cap W(b) = W(c)$. Running again through the argument at the end of the proof of the preceding theorem 3, you deduce from this that F has SAP.

Consider the form $\varphi := \langle -1, a, b, ab\rangle$. It is immediately verified that $\sigma(\varphi) = \pm 2$ for all σ in X. By HMP there exists some $m \geq 1$ such that

$$2^m \times \varphi = 2^m \times \langle -1\rangle \perp 2^m \times \langle a, b, ab\rangle$$

is isotropic. Thus there exists some t in F* such that -t is represented by $2^m \times \langle -1\rangle$ and t is represented by $2^m \times \langle a\rangle \perp \langle b\rangle \otimes [2^m \times \langle 1,a\rangle]$. We further have a decomposition $t = s + bc$ with elements $s \neq 0$, $c \neq 0$, *) s being represented by $2^m \times \langle a\rangle$ and c represented by $2^m \times \langle 1,a\rangle$. Clearly $W(s) = W(a)$ and $W(c) \subset W(a)$. On the other hand $\sigma(t) = 1$ for all σ in X. Thus $\sigma(bc) = -1$ implies $\sigma(s) = +1$ and vice versa, i.e. $W(bc) = W(-s) = W(-a)$. If $\sigma(c) = -1$ then $\sigma(a) = -1$ and $\sigma(bc) = +1$, hence $\sigma(b) = -1$. Thus $W(c) \subset W(a) \cap W(b)$. On the other hand, if $\sigma(a) = \sigma(b) = -1$ then $\sigma(bc) = +1$, hence $\sigma(c) = -1$. Thus actually $W(c) = W(a) \cap W(b)$. This finishes the proof of Theorem 3.22.

Epilogue

In this section we discussed the case that the ordering are "as independent as possible" from each other, while in the preceding section we studied the other extremum, that the orderings are "as dependent as possible" from each other (and F is pythagorean).

To get a better understanding of these cases and of the space of orderings X in general one has to use deeper methods. There is a close connection between X and the set of Krull valuations of F which have a residue class field imbeddable into $\mathbb{R}$. In particular L.Bröcker has generalized the fact that SAP fields have the Hasse-Minkowski property for orderings (Th. 3.22) to a local global principle for weak isotropy for <u>arbitrary fields</u> ([18, Satz 3.9], [53, § 8]). This local-global principle involves beside the real closures of F suitable

---

*) cf. [64, Bemerkung S.39]

henselizations of F. Bröcker's theorem is a powerful tool to analyse the reduced Wittring of a field. In a recent paper [10] Becker and Bröcker in some sense give a complete description of the reduced Wittrings of fields. Rather surprisingly M. Marshall then has succeeded to deduce the essential results of Becker and Bröcker by an abstract approach in the spirit of the present lectures [45], [46], [47], [48]. This abstract approach also covers the structure theory of reduced Wittrings of semilocal rings.

We finally want to emphasize that quadratic form theory as studied here also leads to a better understanding of formally real fields themselves than purely field theoretic methods do, cf. [53] and [8].

References

[1] J.K. Arason and A. Pfister, Beweis des Krullschen Durchschnittssatzes für den Wittring, Invent. Math. 12 (1971), pp. 173-176.

[2] R. Arens and I. Kaplansky, Topological representation of algebras, Trans. Amer. Math. Soc. 63 (1948), pp. 457-481.

[3] C. Arf, Untersuchungen über quadratische Formen in Körpern der Charakteristik 2, J. reine angew. Math. 183 (1941), pp. 148-167.

[4] E. Artin, Über die Zerlegung definiter Funktionen in Quadrate, Abh. Math. Sem. Univ. Hamburg 5 (1927), pp. 100-115.

[5] E. Artin, "Algebraic numbers and algebraic fundtions", Gordon and Breach, New York 1967.

[6] E. Artin and O. Schreier, Algebraische Konstruktion reeller Körper, Abh. Math. Sem. Univ. Hamburg 5 (1926) pp. 85-99.

[7] R. Baeza, "Quadratic forms over Semi-local Rings", Lecture Notes in Mathematics 655, Springer 1978.

[8] E. Becker, "Hereditary pythagorean fields and orderings of higher level", Monografias Mat. N° 29, Inst. Mat. Pura et Applicada, Rio de Janeiro 1978.

[9] E. Becker, K.J. Spitzlay, Zum Satz von Artin-Schreier über die Eindeutigkeit des reellen Abschlusses eines angeordneten Körpers, Comm. Math. Helv. 50, 81-87 (1975).

[10] E. Becker, L. Bröcker, On the description of the reduced Wittring, J. Alg. 52, 328-346 (1978).

[11] A.A. Belskii, Cohomological Witt rings, Math. USSR - Izvestija 2 (1968), pp. 1101-1115.

[12] N. Bourbaki, "Algèbre", Chap. 9, Act. Sc. Ind. 1272, Hermann, Paris, 1959.

[13] N. Bourbaki, "Algèbre", Chap. 5, Act. Sc. Ind. 1102, Hermann, Paris, 1959.

[14] N. Bourbaki, "Algèbre", Chap. 6, Act. Sc. Ind. 1179, Hermann, Paris, 1959.

[15] N. Bourbaki, "Algèbre", Chap. 8, Act. Sc. Ind. 1272, Hermann, Paris, 1959.

[16] N. Bourbaki, "Algèbre commutative", Chaps. 1-2, Act. Sc. Ind. 1290, Hermann, Paris, 1961.

[17] N. Bourbaki, "Algèbre commutative", Chaps. 5-6, Act. Sc. Ind. 1308, Hermann, Paris, 1964.

[18] L. Bröcker, Zur Theorie der quadratischen Formen über formal reellen Körpern, Math. Ann. 210, 233-256 (1974).

[19] L. Bröcker, Characterization of fans and hereditarily pythagorean fields, Math. Z. 151, 149-163 (1976).

[20] L. Bröcker, Über die Anzahl der Anordnungen eines kommutativen Körpers, Arch. d. Math. 29, 458-464 (1977).

[21] R. Brown, Superpythagorean fields, J. Algebra 42, 483-494 (1976)

[22] R. Brown and M. Marshall, The reduced theory of quadratic forms Rocky Mtn. J. of Math., to appear.

[23] H. Cartan and S. Eilenberg, "Homological algebra", Princeton University Press, Princeton, 1956.

[24] G.W.S. Cassels, On the representation of rational functions as sums of squares, Acta Arithmetica 9 (1964), 79-82.

[25] G.W.S. Cassels and A. Fröhlich, "Algebraic number theory", Academic Press, London and New York (1967).

[26] C.M. Cordes, The Witt group and the equivalence of fields with respect to quadratic forms, J. Algebra 26, 400-421 (1973).

[27] T. Craven, The Boolean space of orderings of a field, Trans. AMS 209, 225-235 (1975).

[28] T. Craven, Characterizing reduced Witt rings of fields, J. Alg. 53, 68-77 (1978).

[29] J. Diller, A. Dress, Zur Galoistheorie pythagoräischer Körper, Arch. Math. 16, 148-152 (1965).

[30] A. Dress, The Witt ring as Mackey functor, Notes on the theory of representations of finite groups I, Chap. 2, Appendix A, University of Bielefeld, Bielefeld, 1971.

[31] R. Elman and T.Y. Lam, Quadratic forms over formally real fields and pythagorean fields, Amer. J. Math. 94 (1972), 1155-1194.

[32] D.K. Harrison, "Witt rings", Lecture Notes, Department of Mathematics, University of Kentucky, Lexington, 1970.

[33] N. Jacobson, "Lectures in abstract algebra", vol. III, van Nostrand, 1964.

[34] M. Knebusch, Grothendieck- und Wittringe von nichtausgearteten symmetrischen Bilinearformen, Sitzungsber. Heidelberg, Akad. Wiss. 1969/1970, pp. 93-157.

[35] M. Knebusch, On the uniqueness of real closures and the existence of real places, Com. Math. Helv. 47 (1972), 260-269.

[36] M. Knebusch, A. Rosenberg, R. Ware, Structure of Witt rings and quotients of abelian group rings, Amer. J. Math. 94 (1972), 119-155.

[37] M. Knebusch, A. Rosenberg, R. Ware, Signatures on semi-local rings, J. of Algebra 26 (1973), pp. 208-250.

[38] M. Knebusch and W. Scharlau, Über das Verhalten der Witt-Gruppe bei galoisschen Körpererweiterungen, Math. Ann. 193 (1971) pp. 189-196.

[39] M. Knebusch, "Symmetric bilinear forms over algebraic varieties", Conference on quadratic forms 1976, Queen's papers in pure and appl. math. N° 46, Queen's University, Kingston, Ontario 1977, pp. 103-283.

[40] M. Kula, Fields with prescribed quadratic form schemes, Math. Z. 167, 201-212 (1979).

[41] T.Y. Lam, "The algebraic theory of quadratic forms", Benjamin, Reading, Mass. (1973).

[42] T.Y. Lam, "Ten lectures on quadratic forms over fields", Conference on quadratic forms 1976, Queen's papers in pure and appl. math. N° 46, Queen's University, Kingston, Ontario 1977, pp. 1-102.

[43] J. Leicht and F. Lorenz, Die Primideale des Wittschen Ringes, Invent. Math. 10 (1970), pp. 82-88.

[44] F. Lorenz, "Quadratische Formen über Körpern", Springer Lecture Notes 130, 1970.

[45] M. Marshall, Some local-global principles for formally real fields, Can. J. of Math. 29, 606-614 (1977).

[46] M. Marshall, Spaces of orderings IV, Can. J. of Math. 32, 603-627 (1980).

[47] M. Marshall, The Wittring of a space of orderings, Trans. A.M.S. 258, 505-521 (1980).

[48] M. Marshall, "Abstract Witt Rings", Queen's papers in pure and appl. math. N° 57, Queen's Univ. Kingston, Ontario 1980.

[49] J. Milnor and D. Husemoller, "Symmetric bilinear forms", Springer Verlag, Berlin - Heidelberg - New York, 1973.

[50] J. Milnor, Symmetric inner product spaces in characteristic two, Prospects in Mathematics, Annals of Math. Studies 70 (1970), Princeton University Press, pp. 59-75.

[51] O.T. O'Meara, "Introduction to quadratic forms", Springer Verlag, Berlin - Heidelberg - New York, 1963.

[52] A. Pfister, Quadratische Formen in beliebigen Körpern, Invent. Math. 1 (1966), pp. 116-132.

[53] A. Prestel, "Lectures on formally real fields", Monografias Math. N° 22, Inst. Mat. Pura et Applicada, Rio de Janeiro 1974.

[54] A. Rosenberg and R. Ware, The zero-dimensional Galois cohomology of Witt rings, Invent. Math. 11 (1970), pp. 65-72.

[55] A. Rosenberg, R. Ware, Equivalent topological properties of the space of signatures of a semilocal ring, Publ. Math. Univ. Debrecen 23, 283-289 (1976).

[56] W. Scharlau, Quadratische Formen und Galois-Cohomologie, Invent. Math. 4 (1967), pp. 238-264.

[57] W. Scharlau, Zur Pfisterschen Theorie der quadratischen Formen, Invent. Math. 6 (1969), pp. 327-328.

[58] W. Scharlau, Induction theorems and the structure of the Witt-group, Invent. Math. 11 (1970), pp. 37-44.

[59] W. Scharlau, "Quadratic forms", Queen's papers in pure and applied mathem., N° 22, Queen's Univ., Kingston, Ontario, 1969.

[60] T.A. Springer, Sur les formes quadratique d'indice zero, C.R. Acad. Sci. 234 (1951), pp. 1517-1519.

[61] T.A. Springer, Quadratic forms over fields with a discrete valuation I, Indag. Math. 17 (1955), pp. 352-362.

[62] B.L. van der Waerden, "Algebra I", Heidelberger Taschenbücher Bd. 12, Springer Verlag, Berlin - Heidelberg - New York 1969.

[63] R. Ware, When are Wittrings group rings?, Pacific J. Math. 43, 279-284 (1973).

[64] E. Witt, Theorie der quadratischen Formen in beliebigen Körpern, J. reine angew. Math. 176 (1937), pp. 31-44.

Index